The Digestive & Urinary Systems

Titles in the series

The Muscular & Skeletal Systems

The Circulatory & Respiratory Systems

The Digestive & Urinary Systems

The Reproductive & Nervous Systems

THE HUMAN BODY
& THE ENVIRONMENT
How our surroundings affect our health

The Digestive & Urinary Systems

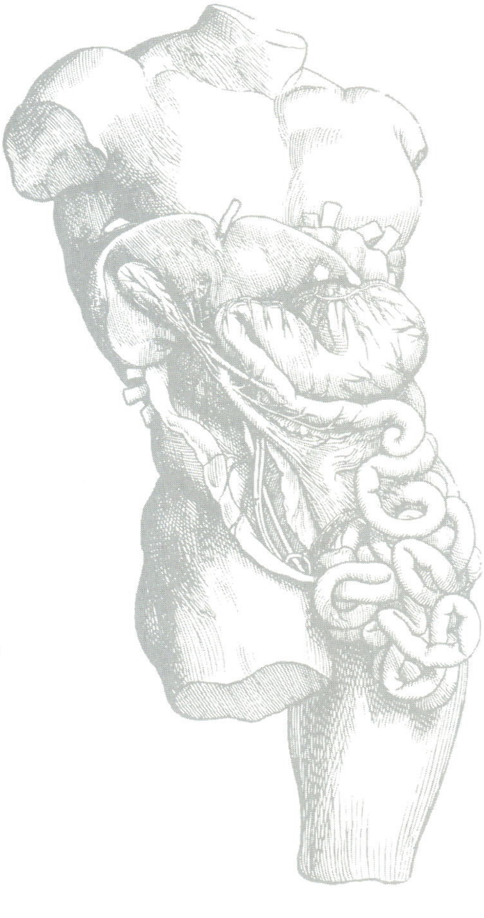

GREENWOOD PRESS
Westport, Connecticut · London

Library of Congress Cataloging-in-Publication Data

Creative Media Applications
The human body & the environment/how our surroundings affect our health
p. cm.—(Middle school reference)
Includes bibliographical references and index.
Contents: v. 1. The muscular & skeletal systems—v. 2. The circulatory & respiratory systems—v. 3. The digestive & urinary systems—v. 4. The reproductive & nervous systems.
ISBN 0-313-32558-8 (set: alk. paper)—0-313-32559-6 (v.1)—0-313-32560-X (v.2) —0-313-32561-8 (v.3)—0-313-32562-6 (v.4)
1. Human physiology—Juvenile literature. [1. Human physiology. 2. Environmental health.] I. Title: Human body and environment. II. Series.
QP37.H8925 2003
612—dc21 2002035217

British Library Cataloguing in Publication Data is available.

Library of Congress Catalog Card Number: 2002035217

ISBN: 0-313-32558-8 (set)
0-313-32559-6 (vol. 1)
0-313-32560-X (vol. 2)
0-313-32561-8 (vol. 3)
0-313-32562-6 (vol. 4)

First published in 2003

Greenwood Press, 88 Post Road West, Westport, CT 06881
An imprint of Greenwood Publishing Group, Inc.
www.greenwood.com

Printed in the United States of America

The paper used in this book complies with the Permanent Paper Standard issued by the National Information Standards Organization (Z39.48–1984).

10 9 8 7 6 5 4 3 2 1

A Creative Media Applications, Inc. Production
WRITER: Robin Doak
DESIGN AND PRODUCTION: Fabia Wargin Design, Inc.
EDITOR: Matt Levine
COPYEDITOR: Laurie Lieb
PROOFREADER: Barbara Francis
ASSOCIATED PRESS PHOTO RESEARCHER: Yvette Reyes
CONSULTANT: Michael Windelspecht

PHOTO CREDITS:
Cover: (*top*) © Chinch Gryniewicz; Ecoscene/CORBIS and (*bottom*) © FoodPix
AP/Wide World Photographs *pages* vi, ix, 9, 10, 15, 21, 23, 29, 35, 38, 44, 51, 54, 55, 56, 64, 67, 72, 76, 80, 88, 107, 111, 113, 115, 118, 121, 124, 125
L.Birmingham/Custom Medical Stock Photo *pages* 3, 4, 62, 69, 102
Articulate Graphics/Custom Medical Stock Photo *page* 5
© Joe Bator/CORBIS *page* 12
Custom Medical Stock Photo *pages* 14, 89, 96
B.Wainwright/Custom Medical Stock Photo *page* 16
Educational Images/Custom Medical Stock Photo *pages* 19, 61
Martin/Custom Medical Stock Photo *page* 25
Barry Bates/Custom Medical Stock Photo *page* 17
Benjamin/Custom Medical Stock Photo *page* 30
SIU/Custom Medical Stock Photo *page* 36
Science Photo Library/Custom Medical Stock Photo *page* 49, 105
K.Glaser/Custom Medical Stock Photo *page* 71
Marka, srl/Custom Medical Stock Photo *page* 75
J.Siebert/Custom Medical Stock Photo *page* 83
© Photodisc *page* 87
M.Bryson/Custom Medical Stock Photo *page* 90
ATC Productions/Custom Medical Stock Photo *page* 92
T.Buck/Custom Medical Stock Photo *page* 93
© Archivo Iconografico, S.A./CORBIS *page* 98
O.J.Staats/Custom Medical Stock Photo *page* 101

ILLUSTRATION CREDITS:
© Lifeart *pages* 1, 6, 7, 63, 95

CONTENTS

INTRODUCTION

LIVING IN A RISKY ENVIRONMENT

You might be surprised at the number of environmental agents that people come in contact with every day. Some of these agents can seriously impact our health. Air and water pollution, for example, can harm the human body. Bacteria, viruses, and fungi that we inhale or ingest can also affect the way that the human body functions.

Environmental health is defined as the human body's reaction to environmental agents, including both natural and human-made substances. In many cases, people have little or no control over the amount of exposure to these environmental agents. Some people, however, choose behaviors that can cause or worsen medical problems. Smoking cigarettes and drinking alcohol, for example, can negatively affect a person's health. Lifestyle decisions made by one person can even affect those around him or her. For example, secondhand cigarette smoke has been proven to have serious affects on the respiratory (RESS-pur-uh-tor-ee) systems of children and adults who are exposed to it.

Just as some lifestyle decisions contribute to ill health, other decisions can help keep the body as healthy as possible. Eating healthy foods and getting plenty of exercise are two important steps that we can take to stay physically fit. Another way to stay healthy is to learn how environmental conditions affect us and how to minimize the damage from environmental agents.

Opposite:
A helicopter sprays pesticides on an orchard near Yakima, Washington. The federal government has limited the use of pesticide sprays in national forests because of the potentially harmful health effects of these chemicals.

THE ENVIRONMENT AND THE BODY

Like other body systems, the digestive, urinary, and lymphatic systems can all be seriously affected by the environment. Pollution, radiation, and chemicals can cause serious medical problems to these systems. These environmental agents have been linked to a number of serious health conditions, including lymphoma, which is cancer of the lymph system, and kidney cancer.

The digestive and urinary systems are particularly affected by the foods and liquids that we put into our bodies. The type of diet that people follow can help or harm these systems. Diets that are high in fats and proteins, for example, can cause digestive and lymphatic problems. On the other hand, diets that are high in fiber can aid digestion and help keep the digestive and urinary systems working properly. Diet choices can also lead to gastritis, constipation, and gastroesophageal reflux disease (GERD). Food-borne diseases are also a problem in the United States. Each year, millions of Americans become sick, sometimes seriously so, after eating food that has been contaminated by bacteria, viruses, and parasites.

Bacteria, viruses, and other organisms can also cause problems with the lymphatic system. These bad bugs can cause swollen lymph glands and lymph vessels, as well as an enlarged spleen. Chemicals can also damage the lymphatic system. Some of these substances may lead to lymphoma.

HOW TO USE THIS BOOK

Each volume of this series covers a number of body systems and the diseases—environmental and otherwise—that can affect them. Each volume is organized to include the basic anatomy of each system, including its structure and function. Volume 3 includes the digestive, urinary, and lymphatic systems.

Nonenvironmental medical conditions and environmental medical conditions for the body systems are covered in separate chapters. Each disease listing includes the condition's causes, symptoms, and treatments or cures. Additional information may include diagnostic tools, statistics, and historical information.

Opposite:
A tourist buys produce at a fruit market in La Colonia Tovar, Venezuela. A healthy diet that includes fresh fruits and vegetables is essential for good overall health and is especially important for good gastro intestinal and urinary tract well-being.

The final chapter in each volume offers tips on how to keep the body healthy. The glossary gives definitions and pronunciations of some more difficult words in the book. For a complete list of pronunciations, however, consult a medical dictionary. The index can be used to find a particular disorder easily.

Finally, the volumes are not meant to be used for self-diagnosis of medical conditions. Those who have health problems should always consult a medical professional.

Note: All metric conversions in this book are approximate.

1

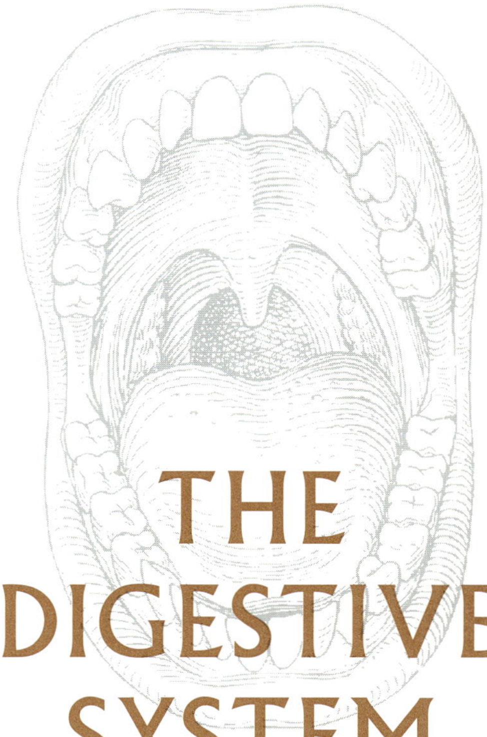

THE DIGESTIVE SYSTEM

All people need good, healthy foods to keep their bodies functioning effectively. Healthy foods contain the vitamins, minerals, and other nutrients that the body needs in order to stay fit and active. How is an apple or carrot broken down and converted into bits that are usable by the body? That is the job of the *digestive* (dye-JESS-tiv) *system.*

The digestive system extends from the mouth to the rectum. Each organ has an important role to play. As food moves through the body, it is broken down into smaller and smaller pieces. Anything that the body can use for nourishment and energy is absorbed into the bloodstream. The bloodstream circulates such substances as vitamins, glucose, lactose, amino acids, and cholesterol (kuh-LESS-tur-awl) to tissues and organs throughout the body. (*Cholesterol* is a soft, waxy substance found in the blood and cells of the body. It helps the body produce cell membranes, hormones, and tissues.) Unusable food particles are eventually expelled from the body as waste. From start to finish, the entire digestive process can take anywhere from twelve hours to two days.

The digestive system is made up of digestive and accessory organs. The digestive organs are those that take part in the actual digestion of food. Digestive organs include the mouth, esophagus, stomach, small intestine, and large intestine. The accessory organs are those that help in the digestive process, but are not part of the digestive *tract,* or system. Accessory organs include the teeth, tongue, salivary glands, pancreas, liver, and gallbladder.

DOWN THE HATCH

The first digestive organ involved in turning food into energy is the mouth. Digestion begins when a person places food into the mouth and begins to chew. Inside the mouth, accessory organs go to work on the food. Teeth break down the food into smaller pieces. The tongue mixes the food and pushes it between the teeth for more mashing. During chewing, saliva (suh-LYE-vuh) is secreted from the mouth's salivary (SAL-ih-veh-ree) glands. Enzymes in the saliva also help break down food.

> **Fast Fact**
>
> The wavelike contractions of the muscles of the esophagus, stomach, and small intestine are known as *peristalsis* (perr-ih-STAL-siss).

After the mouth has done its job, food passes into the second digestive organ, the *esophagus* (ih-SAHF-ih-guss). The esophagus is a long, narrow tube that connects the mouth to the stomach. Muscles in the esophagus contract in wavelike motions, propelling the food downward. From the esophagus, the food passes into the stomach. A valve, called a sphincter, closes to make sure that the food remains in the stomach. The sphincter valve also protects the esophagus from stomach fluids.

In the stomach, food is churned and mixed with the stomach fluids. These fluids are made up of acids and enzymes that break down the food pieces even further. Muscles lining the stomach mix and mash the food, breaking up fats and proteins. Food can remain here for just a few minutes or for several hours.

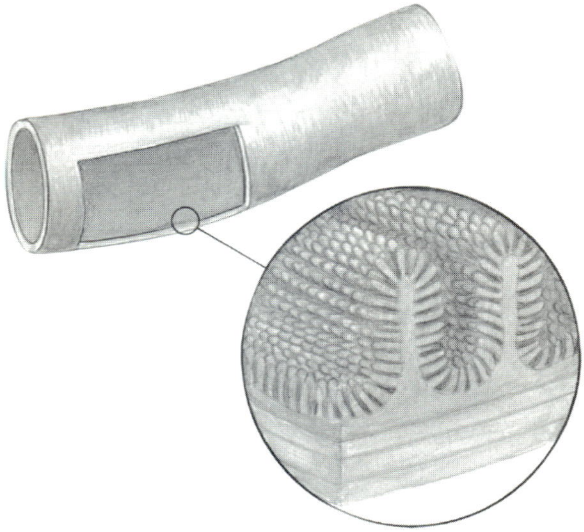

This illustration shows a section of intestine and an enlargement of the villi that line the intestines and absorb nutrients from small food particles.

Once the food inside the stomach has been mashed and smashed, it is pushed slowly into the *small intestine* (in-TESS-tin). The small intestine is a coiled tube that measures 16 to 20 feet (4.8 to 6 meters) in length. As food passes through this tube, it is broken down into tiny, molecule-sized pieces. Small, fingerlike organs that line the intestine, called *villi* (VIL-lye), absorb nutrients from the tiny food pieces. The villi pass these nutrients into the bloodstream and lymphatic system to be circulated throughout the body.

Once the villi of the small intestine have removed all the nutrients and usable materials from the food, anything left over is known

> **Fast Fact**
>
> Every square inch of the small intestine is lined with billions of villi.

as waste. Waste may consist of fiber, fat, and old cells. The waste leaves the small intestine and enters the *large intestine.* The large intestine is only about 5 feet (1.5 meters) long—shorter but much wider than the small intestine. Here, any remaining minerals and extra water are removed from the waste. As waste moves through the large intestine, it becomes more solid. Eventually, it is expelled from the rectum, the final part of the large intestine. The rectum ends at the anus, the opening that allows waste to exit the body.

ACCESSORY ORGANS

Although the liver, gallbladder, and pancreas are not part of the digestive tract, they are extremely important to the digestive process. The liver, for example, performs several digestive functions. This reddish-brown, cone-shaped organ does the following:

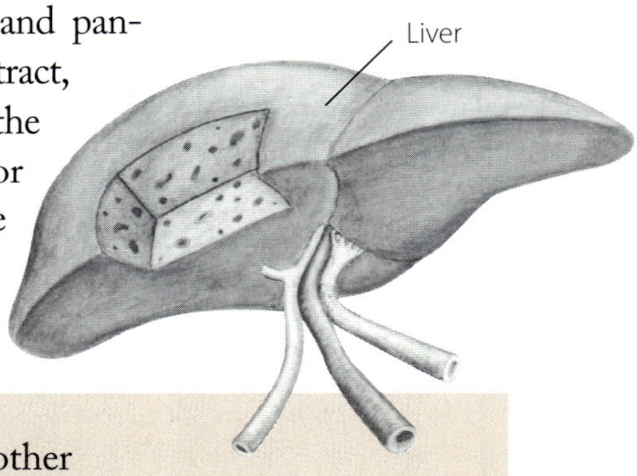

- stores vitamins, minerals, and other nutrients that the body gets from food
- purifies the blood from the stomach and intestines as it passes through by breaking down substances and removing waste from the body
- detoxifies harmful substances, including chemicals and poisons
- helps regulate the amount of some chemicals in the body
- makes bile, a substance that breaks down fats and removes toxins from foods in the small intestine; other substances made in the liver include bilirubin (bil-ih-ROO-bin), albumin (AL-byoo-min), and enzymes (Bilirubin is a reddish-yellow pigment found in blood, bile, and urine. Albumin is a protein found in blood and muscles.)

Another important accessory organ is the *gallbladder.* Bile, made in the liver, is stored here. When bile is needed, the gallbladder contracts, squeezing the bile into the small intestine to start its work.

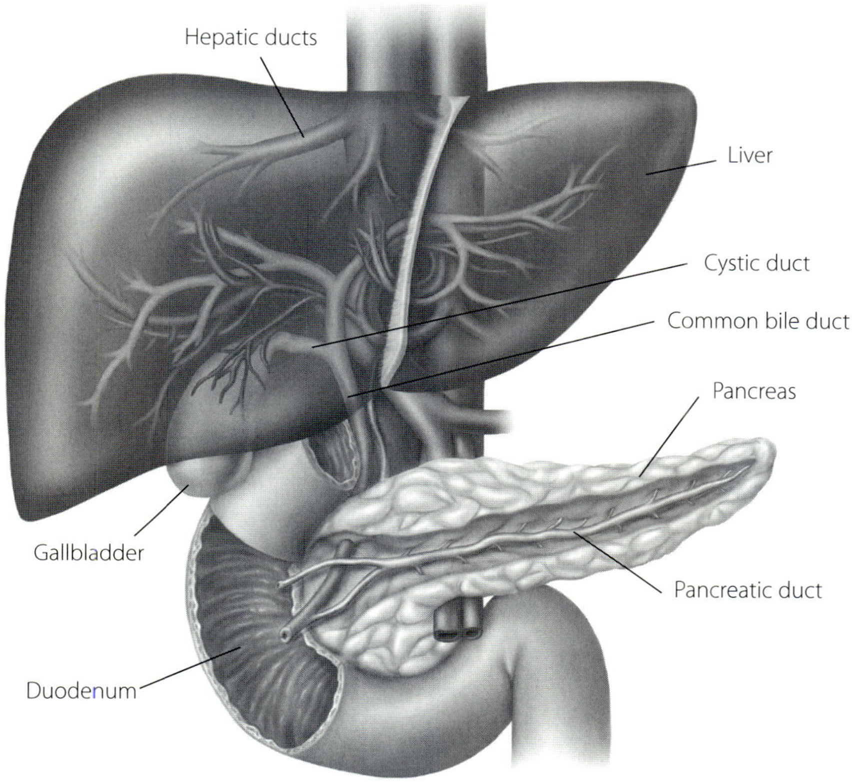

The positions of the gallbladder, liver, and pancreas within the abdomen and the parts of each organ are illustrated in this drawing.

The *pancreas* (PAN-kree-uss) is a small accessory organ that is connected to the small intestine. It produces more than twenty-five different enzymes that enter the small intestine to break down the carbohydrates, fats, and proteins that are present in food.

Fascinating Liver Facts

- The liver is the only organ in the body that can regenerate itself.
- The liver can store about 1 pint (0.5 liters) of blood.
- The liver can lose as many as three-fourths of its cells before it stops working.

Parts of the Digestive System

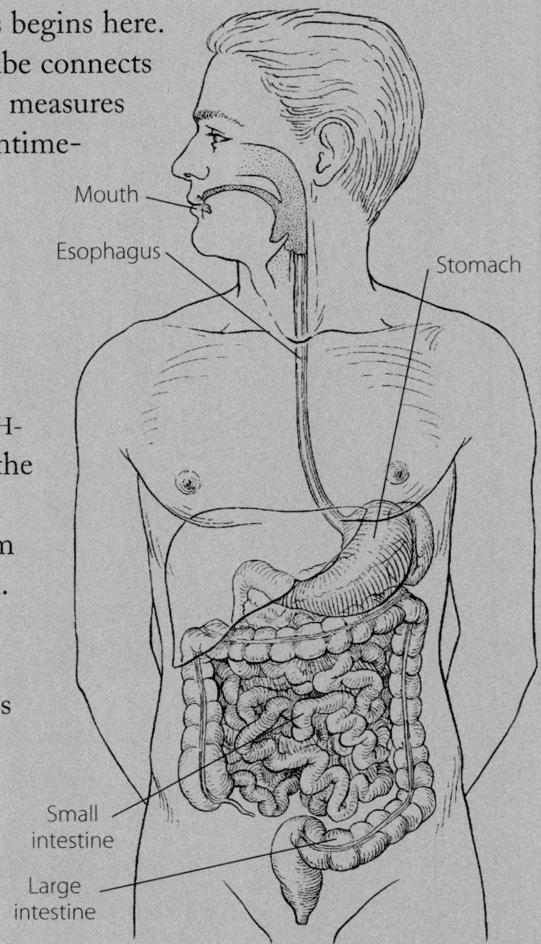

MOUTH—The digestive process begins here.

ESOPHAGUS—This muscular tube connects the throat to the stomach. It measures 10 to 13 inches (25 to 33 centimeters) in length.

STOMACH—This pouchlike organ receives, holds, and partly digests food.

SMALL INTESTINE—The small intestine has three different parts: The duodenum (doo-AH-den-um) is the upper section, the jejunum (jeh-JOO-num) is the central section, and the ileum (IL-ee-um) is the lower section.

LARGE INTESTINE—The large intestine has three different parts: The cecum (SEEK-um) is the upper section, the colon (KOH-lun) is the central section, and the rectum is the lower section. The appendix, a branch of the large intestine, has no known digestive function.

LIVER—This important organ makes bile and helps the body use food.

GALLBLADDER—Bile is stored here until it is needed by the body.

PANCREAS—Enzymes that break down food are produced here.

2

DIGESTIVE DISORDERS

Health conditions can affect any part of the digestive system, from the esophagus to the rectum. Disorders of the digestive system are common in the United States. As many as 62 million Americans may develop such disorders each year. In 2000, digestive-system problems accounted for more than 41 million visits to a doctor's office or hospital outpatient clinic.

Over time, the digestive system suffers wear and tear, just like other body systems. As people age, their livers produce fewer enzymes and proteins. For this reason, some digestive system disorders develop as a person ages.

CONGENITAL DISORDERS

A number of digestive-system problems are congenital, or present at birth. These problems can be a result of abnormal genes, inherited from one or both parents. They may also be the result of a birth defect, a problem that occurs as the fetus develops within its mother's womb.

ALAGILLE SYNDROME

Alagille (ah-la-ZHEEL) *syndrome* is a rare, inherited disorder that occurs during the first year of an infant's life. The infant loses bile ducts inside the liver, and the bile ducts on the outside of the liver narrow. Symptoms of Alagille syndrome include jaundice, diarrhea, and failure to grow during the first months of life. Children born with this disorder may also have serious heart, lung, or kidney problems. Doctors treat Alagille syndrome with medicines and vitamin supplements. In severe cases, patients may require a liver transplant.

ALPHA-1 ANTITRYPSIN DEFICIENCY

Antitrypsin (an-tee-TRIP-sin) *deficiency* is the most common inherited liver problem in newborns. Antitrypsin is a protein that is made in the liver. This important substance protects the lungs from damage. Without antitrypsin, the lungs cannot function properly. According to the National Library of Medicine, three out of every four adults with a severe case of this disorder will eventually develop *emphysema* (em-fih-SEE-muh), a serious lung condition. For this reason, people with the disorder should avoid smoking. Antitrypsin deficiency can also lead to such liver conditions as cirrhosis and hepatitis.

Symptoms of antitrypsin deficiency include shortness of breath, wheezing, and unexplained weight loss. If the patient also has emphysema, cirrhosis, or hepatitis, symptoms of these diseases will also be present. Doctors treat this condition with transfusions of purified antitrypsin from donor blood plasma. A *transfusion* is a medical procedure in which blood or another liquid is transferred into an artery or a vein. People with antitrypsin deficiency may need to have a transfusion every week. Those with severe liver damage as a result of cirrhosis may require a liver transplant.

Mary Pierce carried the Olympic torch in Washington, D.C., on behalf of the American Lung Association before the 2002 Winter Olympics. Pierce suffers from a hereditary lung and liver disease that resulted in a double lung transplant eight years earlier.

Meghan Nadeau hands off the baton in the 100-meter relay at her high school in Ticonderoga, New York. Meghan suffers from cystic fibrosis, which has weakened her lungs but has not stopped her from running track.

CYSTIC FIBROSIS

Although most commonly thought of as a respiratory disease, cystic fibrosis (SISS-tik fye-BROH-siss), or CF, can also have a devastating effect on the digestive system. CF is a fatal, hereditary condition in which mucous membranes in the respiratory tract produce abnormally thick mucus. Even before a baby is born, the mucus builds up in the tubes that connect the pancreas to the small intestine. It blocks enzymes that digest food from entering the intestines. As a result, the pancreas becomes damaged and scarred. This can result in failure of the pancreas, improper absorption of nutrients in the intestines, and malnutrition.

CF is a *progressive* condition. This means that it gets worse over time. In the past, CF was always fatal in childhood. Today, thanks to better medicines and other treatments, children born with CF are living longer. Children with pancreatic damage, for example, can take enzyme supplements to aid the digestive system. Today, the life expectancy for a person with CF is about thirty years. Although the effects on the digestive system can be serious, people who die as a result of the condition usually die of lung disease.

GALACTOSEMIA

Children with *galactosemia* (guh-lak-toh-SEEM-ee-uh) are born without a certain liver enzyme that breaks down galactose. Galactose is a sugar that is present in milk. When babies born with this condition are fed milk, problems result, including vomiting, jaundice, and convulsions. Too much galactose in the body can lead to damage of the liver, kidneys, brain, and eyes.

To prevent infants from suffering serious damage as a result of galactosemia, some states test all newborn infants for the condition. In other states, infants are discovered to have the disorder after they begin exhibiting such symptoms as jaundice, vomiting, refusal to eat, poor weight gain, irritability, and convulsions. The disease is diagnosed through blood tests.

If a baby is found to have the disorder, parents must immediately stop feeding the child milk. Instead, the infant must be fed soy-based or other nursing formulas. Because there is no cure for the disease, people with galactosemia must avoid drinking milk or eating foods made with milk products for the rest of their lives.

HIRSCHSPRUNG'S DISEASE

Hirschsprung's disease (HD) is a genetic condition in which an infant is born with an abnormal large intestine. The nerve cells on the colon are absent, which affects muscle function there. Without muscle movement, waste builds up in the large intestine, leading to infection, a burst colon, and death.

The severity of HD depends upon how much of the intestine is without nerve cells. In the most serious cases, an infant is born missing nerve cells in the entire large intestine, as well as part of the small intestine. Infants with this type of HD require immediate medical attention.

Symptoms of HD include constipation, vomiting of green bile, gas, and abdominal discomfort. Patients suffering from this condition are treated with "pull-through" surgery. During this procedure, the abnormal section of the intestine is removed. The part of the intestine that works properly is then connected to the anus. The earlier the condition is detected, the better the chance for completely curing HD through surgery.

INTESTINAL MALROTATION

Babies born with this condition have intestines that don't coil properly in the abdomen. Although not necessarily a serious condition, malrotation may lead to other, more serious, medical conditions that can create intestinal blockages or restrict the flow of blood into the intestines. In fact, malrotation may not be diagnosed until a more serious health problem arises.

Symptoms of malrotation include abdominal pain, cramping, vomiting, and diarrhea or bloody stools. Children born with this condition usually require surgery to correct the problem.

JAUNDICE

Jaundice is a condition in which bilirubin builds up inside the body. As a result, the skin, eyes, and mucous membranes begin to take on a yellowish color. (*Mucous membranes* are layers of mucus-secreting cells that line the tubes and cavities of the body.) Jaundice often affects newborn infants whose livers have not yet developed completely. Newborn jaundice usually lasts just a few days.

When adults develop jaundice, however, it is usually an indication of a liver problem. Bilirubin buildup can indicate a serious condition, such as hepatitis, gallstones, or cancer of the liver or pancreas. As a result, people with jaundice should seek medical evaluation immediately.

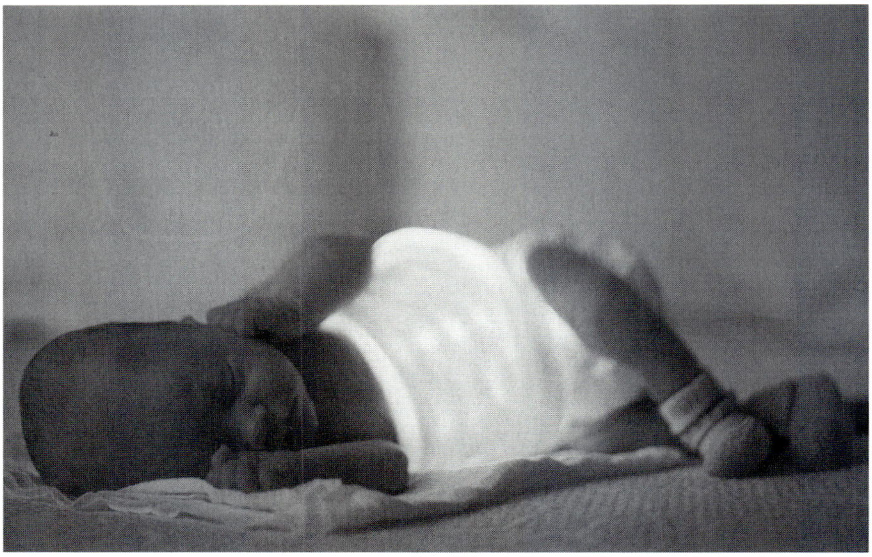

A newborn is bathed in purple ultraviolet light as a treatment for jaundice.

One common disease that causes jaundice is Gilbert's syndrome. This inherited condition upsets bilirubin production in the liver. Many people with this condition will never have any symptoms and appear jaundiced only after exercise, sickness, or stress. The disease is not serious and treatment is usually not necessary.

MORE CONGENITAL LIVER DISORDERS

There are other congenital disorders that affect the liver. They include the following:

DISEASE	Hemochromatosis (hee-moh-kroh-muh-TOH-siss)
NATURE OF DEFECT	The liver absorbs and stores too much iron.
SYMPTOMS	Joint pain, fatigue, weight loss, darkening of the skin, abdominal pain, and enlargement of the liver; complications include heart and skin damage, cirrhosis, and liver cancer
TREATMENT	Regular removal of blood from a patient's body in order to decrease the amount of iron that is present
DISEASE	Porphyria (por-FIR-ee-uh)
NATURE OF DEFECT	The liver makes too much porphyrin (POR-fih-rin), a chemical that helps to make hemoglobin. Hemoglobin (HEE-moh-gloh-bin) is the substance that gives blood its red color and carries oxygen.
SYMPTOMS	Abdominal pain, cramps, seizures, weakness, hallucinations and other mental disorders (nervous system damage), blisters, swelling, itching, and sensitivity to the sun (skin damage)
TREATMENT	Medicines or a diet that is high in carbohydrates
DISEASE	Wilson disease
NATURE OF DEFECT	Copper builds up in the liver.
SYMPTOMS	Jaundice, swelling of the abdomen, vomiting blood, and enlargement of the spleen; complications include liver disease, permanent liver damage, hepatitis, cirrhosis, liver failure, nervous system damage, brain damage, and eye damage
TREATMENT	Medicines or a diet that is low in copper

AUTOIMMUNE DISORDERS

Autoimmune (aw-toh-im-MYOON) *disorders* occur when the body's immune system does not function properly and attacks healthy cells, tissue, and organs in the body. An autoimmune disorder can affect any body system, including the digestive system.

AUTOIMMUNE HEPATITIS

This type of hepatitis (hep-uh-TYE-tiss) occurs when the autoimmune system mistakenly attacks the body's liver cells. The liver becomes swollen and may become permanently damaged. Liver damage can result in cirrhosis. Autoimmune hepatitis most often affects females between the ages of fifteen and forty.

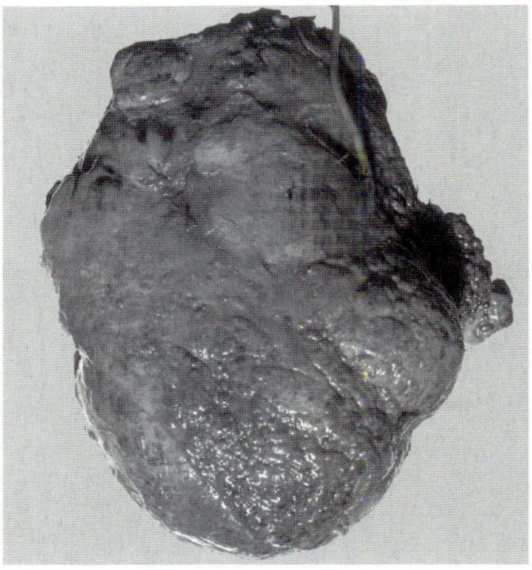

This swollen, cirrhosis-damaged liver may have been the result of hepatitis.

Symptoms of this condition include an enlarged liver, jaundice, skin rashes, itching, and pain in the joints and abdomen. If diagnosed early enough, autoimmune hepatitis can be treated effectively with medicines that control the immune system. Some patients may have to take these medicines for the rest of their lives. However, patients whose livers have been severely damaged as a result of the condition may require a liver transplant.

CELIAC DISEASE

When people with *celiac* (SEE-lee-ak) *disease* eat gluten, a protein found in wheat, rye, oats, and other grains, it damages their small intestines. Many doctors believe that the presence of gluten in the body of people with this disease causes the autoimmune system to attack and damage the small intestine. The villi that line the small intestine lose their ability to effectively absorb nutrients. As a result, people suffering from celiac disease may experience anemia (uh-NEE-mee-uh), which is a lack of red blood cells, weight loss, and vitamin deficiencies. In turn, these deficiencies can lead to bone and tooth problems, including rickets and osteomalacia (ah-stee-oh-muh-LAY-shuh), or softening of the bones.

Symptoms of celiac disease include abdominal pain and bloating, diarrhea, cramps, fatigue, and pain in the joints and bones. People with this condition must avoid food products that contain gluten for the rest of their lives. There are also medicines that treat the condition.

CROHN'S DISEASE

Crohn's disease is an inflammation of the small intestine. The disorder, also called enteritis (en-tih-RYE-tiss) or inflammatory bowel disease, can also affect any other part of the digestive system. It most commonly affects the ileum, the end section of the small intestine.

This woman has Crohn's disease and blames environmental exposure to chromium in her hometown for causing her illness. She holds photographs of her sister and brother who died as a result of the disease.

Crohn's disease is most often an autoimmune disorder. The disease tends to run in some families, however, and researchers have found that smokers and former smokers are more likely to develop the condition. Symptoms include pain in the abdomen, diarrhea, fever, weight loss, and bleeding from the anus. Crohn's disease can lead to other, more serious conditions, including anemia, ulcers, blockage of the intestines, arthritis (arth-RYE-tiss), gallstones, and liver disease.

Doctors treat the condition with anti-inflammatory medicines, dietary changes, supplements, or surgery to remove the damaged section of the intestine. There is currently no cure for Crohn's disease.

MISCELLANEOUS CONDITIONS

There are many digestive diseases that have unknown causes. Researchers continue to work to understand more about these conditions.

INTESTINAL DISORDERS

Several disorders with unknown causes affect the intestines. These include appendicitis, colorectal polyps, irritable bowel syndrome, and ulcerative colitis.

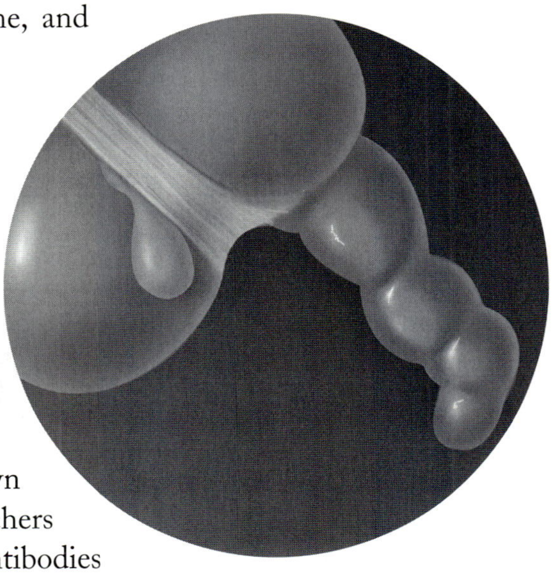

This illustration shows an enlarged appendix.

APPENDICITIS

Appendicitis (uh-pen-dih-SYE-tiss) occurs when the appendix becomes swollen and inflamed. The appendix is a small pouch that protrudes from the right side of the colon. The appendix has no known digestive function, but researchers believe that it may produce antibodies that fight disease. In some cases, the appendix bursts, resulting in infections and death. People between the ages of ten and thirty are most often affected by this condition. According to the National Library of Medicine, appendicitis is one of the most common reasons for emergency surgery performed on children in the United States.

Doctors are not sure what causes appendicitis. Symptoms of the condition include sharp pain in the abdomen that starts near the belly button and then moves to the right side; nausea; loss of appetite; vomiting; diarrhea or constipation; fever; and abdominal swelling. People with these symptoms should contact a doctor immediately. Once the appendix has ruptured, abdominal pain may ease for a while. The pain will soon return, however, as the lining of the abdomen becomes inflamed. This condition, called peritonitis, can cause death if not treated promptly.

Most cases of appendicitis are treated with surgery and antibiotics. Surgery to remove the appendix is known as an *appendectomy* (ap-en-DEK-tuh-mee). To reach the appendix, surgeons make a small incision in the abdomen. If the appendix has ruptured, doctors will also wash the inside of the abdomen and leave a small tube there for pus and other fluids to drain. Because the appendix has no known function, it is not a problem for a person to continue a normal healthy life without one.

COLORECTAL POLYPS

Colorectal polyps (koh-loh-REK-tul PAHL-ipz) are growths of extra tissue on the inside of the intestines or the colon that protrude into the abdomen. Polyps are a common health condition, affecting as many as one out of every five adult Americans. People over the age of fifty who have a family history of this condition are at higher risk than other people are. In addition, those who smoke, drink alcohol, eat a high-fat diet, and are overweight are at a higher risk for colon polyps.

Symptoms of one or more colorectal polyps include bleeding from the anus, blood in the stool, and diarrhea or constipation for a week or longer. People with any of these symptoms should seek medical attention immediately. Although most colon polyps are not harmful, certain types may become cancerous over time. Doctors will remove polyps that are more than 1 centimeter (0.4 inches) in size and |check them under a microscope to make sure that they are not cancerous.

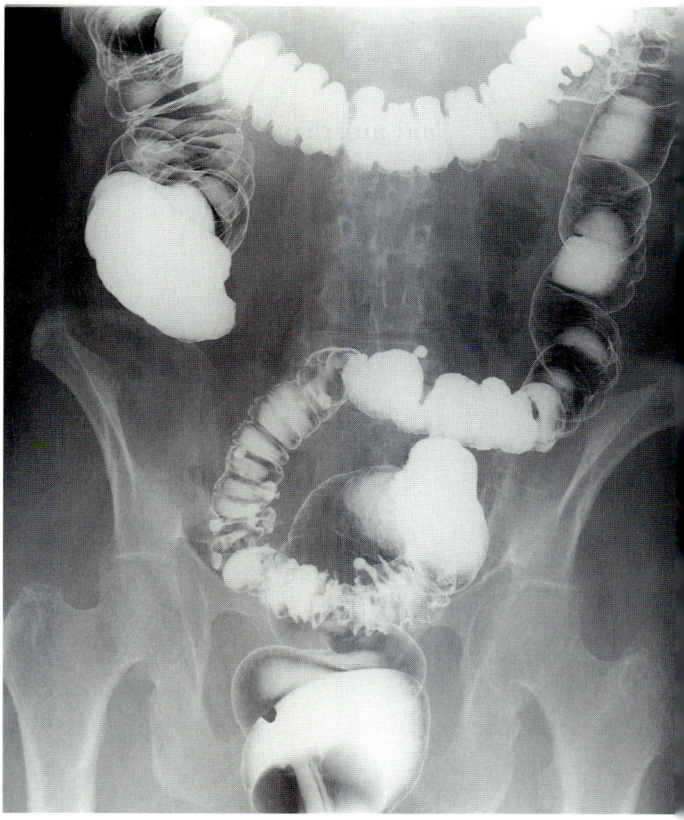

The winding path of the colon through the abdomen is shown in this X-ray photograph.

IRRITABLE BOWEL SYNDROME

Irritable bowel syndrome (IBS) is an intestinal condition that occurs when the muscles in the intestines move more often than is normal. It is known by a variety of other names, including spastic colon, nervous indigestion, and functional colitis. This common condition is not dangerous, but it can cause extreme discomfort and lead to disability. The disease most commonly affects women and usually begins between the ages of twenty and thirty.

Symptoms of the condition include cramps, gas, and bloating. Although no cause is known, doctors believe that stress can make IBS worse. People who suffer from IBS may have colons that are much more sensitive than other people's. To treat this condition, doctors recommend eating a healthy diet with plenty of high-fiber foods, as well as eating small portions at mealtime. People with IBS should also avoid any foods that have triggered the condition in the past.

ULCERATIVE COLITIS

Ulcerative colitis (koh-LYE-tiss), sometimes known simply as colitis, occurs when the large intestine becomes inflamed, and *ulcers,* or sores, develop on its outer lining. Patients have periodic attacks of the disease, triggered by such factors as stress or illness. These attacks eventually cause the intestinal lining to become scarred and thickened.

Although doctors cannot be sure what causes this condition, some think that it may be an autoimmune disorder. Others think that it may be an inherited disease, because it tends to run in some families. People between the ages of fifteen and forty are more likely than others to develop ulcerative colitis.

Symptoms of this disease include fatigue, loss of appetite, weight loss, and rectal bleeding. The disease can lead to other health problems, including liver disease, kidney stones, anemia, arthritis, and skin rashes. An estimated 5 percent of all patients with the condition later develop colon cancer. Colitis is a chronic disease for which there is no current cure. Doctors treat the disease with medicines, but in severe cases, the colon must be surgically removed.

PLUMMER-VINSON SYNDROME

Plummer-Vinson syndrome, also known as esophageal (ih-sahf-ih-JEE-ul) web, is a condition that affects the esophagus. Webs of tissue grow throughout the esophagus, interfering with food digestion. Although the exact cause of this condition is not known, researchers have linked it to chronic iron-deficiency anemia. Genetic factors may also play a role. This condition can increase the risk for cancer of the esophagus. Doctors treat Plummer-Vinson syndrome by prescribing iron supplements which can ease a patient's symptoms.

GALLSTONES

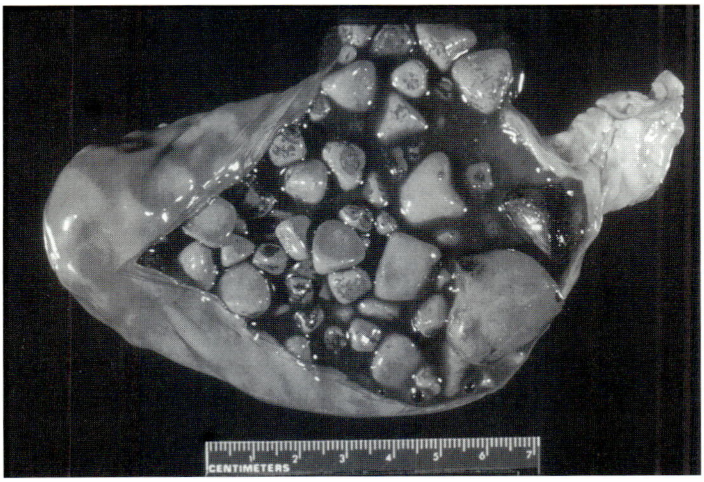

This cross-section of a gallbladder shows the effect of multiple gallstones.

Gallstones, also known as cholelithiasis (koh-leh-lih-THYE-uh-siss), are rocklike pieces that form in the gallbladder when bile hardens. Gallstones affect 20 million Americans—20 percent of all U.S. women and 10 percent of all U.S. men, according to the National Institute for Diabetes and Digestive and Kidney Diseases (NIDDK). Gallstones vary in size from tiny formations that are no larger than a grain of sand to huge stones the size of golf balls. A person may develop just one gallstone or many.

There are two types of gallstones: cholesterol stones and pigment stones. *Cholesterol stones,* yellowish-green in color, are made up mostly of hardened cholesterol found in bile. Cholesterol stones are the most common type of gallstone, making up 80 percent of all gallstones. Doctors believe that cholesterol stones result when there is an excess of cholesterol in the body or when the gallbladder does not empty properly.

Pigment stones, small and dark in color, are made up mostly of bilirubin found in bile. Doctors are less sure what causes this type of gallstone. Many people who develop pigment stones have other health conditions, including cirrhosis, biliary (BIL-ee-eh-ree) tract infections, and sickle-cell anemia.

When gallstones block the path of bile into the stomach, serious problems can result. They can cause the gallbladder and the bile ducts to become inflamed. If the blockage is not removed, it can cause infection and permanent damage to the gallbladder, liver, and pancreas.

Some people with gallstones never experience any symptoms and do not require medical treatment. These gallstones are nicknamed "silent stones." Other people experience severe pain in the upper right abdomen, between the shoulder blades, or under the right shoulder. Other symptoms include indigestion, nausea, fever, chills, and jaundice. People with these symptoms should see a doctor immediately.

Fast Fact

The gallbladder is not a vital organ. A human body can function without it. After removing the gallbladder from a patient, surgeons connect the liver to the small intestine. That way, bile flows directly from the former into the latter.

Doctors can treat some gallstones with medicines that dissolve them. Sometimes, the stones can be broken into smaller pieces by shock waves directed at the gallbladder. Smaller pieces will not obstruct the bile ducts and cause blockages. If these methods are unsuccessful, surgery to remove the gallbladder may be performed. Each year, more than 500,000 people have their gallbladders removed during a procedure called a *laproscopic cholecystectomy* (lap-uh-roh-SKAH-pik koh-luh-siss-TEK-tuh-mee). During this procedure, doctors make a small incision in the abdomen and remove the organ.

Certain factors, including obesity, pregnancy, and diabetes, place some people at high risk of developing gallstones. *Diabetes* is a chronic disease in which levels of sugar in the blood are higher than normal. Native Americans and Mexican Americans are at higher risk for gallstones, as are people over the age of sixty.

STOMACH DISORDERS

Two conditions with unknown causes that affect the stomach are hiatal hernias and pyloric stenosis.

HIATAL HERNIA

This professional football player is recovering from hernia surgery. His condition was most likely caused by excessive strain while playing football.

Hernias are ruptures or tears in the lining of a muscle, tissue, or membrane that holds an organ in place. A *hiatal* (hye-AY-tul) *hernia* is a condition in which the stomach protrudes up into the chest through an opening in the diaphragm called the *esophageal hiatus.* The *diaphragm* (DYE-uh-fram) is the sheetlike muscle that separates the stomach from the chest. Hiatal hernias are common conditions in the United States. They most often occur in people over the age of fifty.

> **Fast Fact**
>
> There are many different types of hernias. The most common type is the *inguinal* (IN-gwin-ul), or groin hernia, which occurs when part of the intestine pokes through the abdominal wall in the groin area.

Although doctors are not sure exactly what causes a person to develop a hiatal hernia, they believe that part of the problem is increasing age. As a person gets older, the hiatus can stretch and weaken. Obesity and smoking also seem to increase the risk of getting this condition. In addition, hiatal hernias may be caused by sudden straining, including coughing, vomiting, or physical exertion.

Most people with hiatal hernias do not show any symptoms. Some people, however, experience heartburn. A hiatal hernia generally does not need to be treated unless a person experiences pain or constant heartburn. People with hiatal hernias can control the unpleasant symptoms by eating smaller meals, avoiding foods that may cause heartburn, avoiding tight clothing, and maintaining a healthy weight. Sometimes, however, a hernia may become very large, or it may twist and cut off the supply of blood to the stomach. In these rare and extreme cases, doctors will perform surgery.

PYLORIC STENOSIS

Pyloric stenosis (pye-LOR-ik sten-OH-siss) is a condition in which the muscles of the lower stomach thicken and the opening to the intestines narrows. As a result, food is unable to pass from the stomach to the intestines. Pyloric stenosis commonly affects infants in the United States. According to the National Library of Medicine, about one out of every 4,000 babies is born with the condition.

Doctors don't know the exact cause of the disease. It may be a congenital disorder or an autoimmune condition. Babies with the disease usually begin to display symptoms when they are about three weeks old. These symptoms include vomiting, constipation, failure to gain weight, and dehydration. Doctors treat pyloric stenosis with surgery to widen the narrowed stomach opening. This surgery usually completely cures the condition.

LIVER DISORDERS

There are a number of conditions with unknown causes that can affect the liver. Two of these disorders are primary biliary cirrhosis and sclerosing cholangitis.

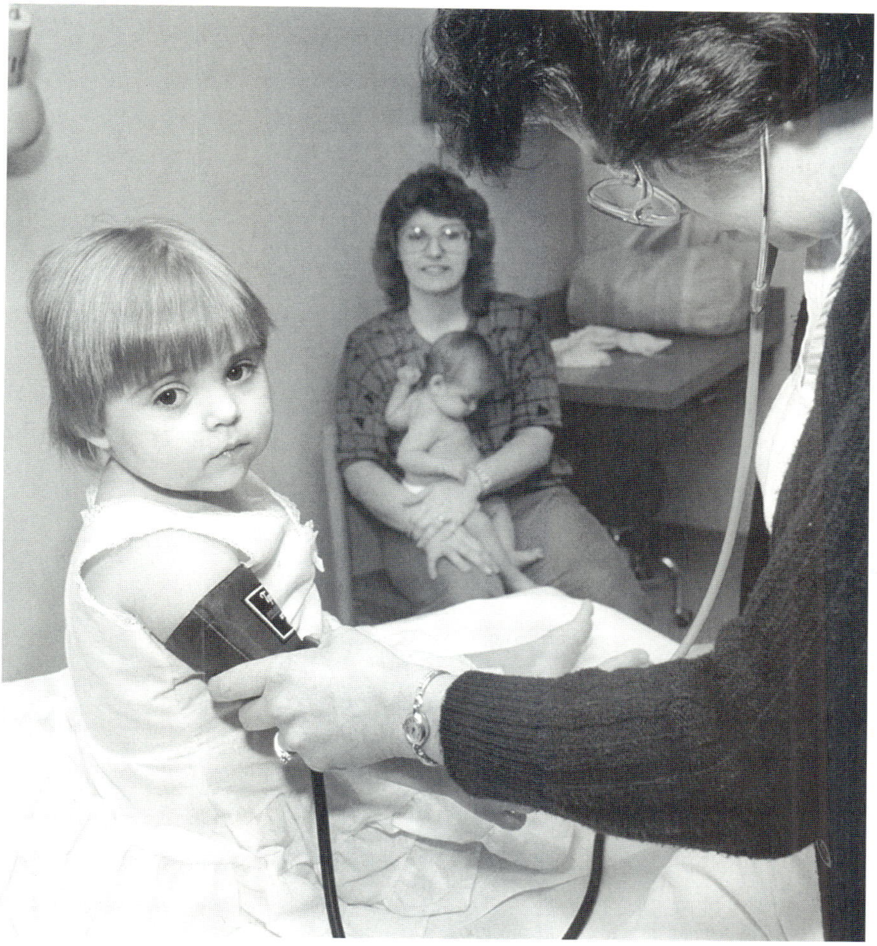

Kim Anthis has her blood pressure checked by a doctor while her sister and mother watch. Kim and her sister suffer from biliary atresia, a liver-scarring disease. Her sister died from the disease at age three, but Kim recently graduated from high school and lives an active life.

PRIMARY BILIARY CIRRHOSIS

Primary biliary cirrhosis (sih-ROH-siss) occurs when the bile ducts become damaged or destroyed, causing a buildup of bile in the liver. Eventually, the condition can lead to cirrhosis of the liver. Doctors are not sure what causes primary biliary cirrhosis, but some researchers believe that it may be an autoimmune disorder. The condition most often affects women between the ages of thirty and sixty.

Symptoms of the condition include itchy skin, fatigue, jaundice, cholesterol deposits under the skin, and fluid retention. Doctors treat primary biliary cirrhosis by prescribing calcium and vitamin supplements, as well as medicines to ease the symptoms. If left untreated, the disease can permanently damage the liver, and a liver transplant is then necessary.

SCLEROSING CHOLANGITIS

Sclerosing cholangitis (skluh-ROH-sing koh-lan-JYE-tiss) is a condition in which the bile ducts inside and outside the liver become inflamed and scarred. The ducts become blocked and bile clogs the liver. As a result, patients may develop primary biliary cirrhosis or cirrhosis of the liver. The final result can be permanent liver damage and the need for a transplant.

Doctors aren't sure why some people develop sclerosing cholangitis. Some researchers believe that it may be caused by a bacteria or virus. The condition usually affects men between the ages of thirty and sixty. Symptoms include itching, jaundice, fatigue, and an enlarged liver or spleen. People with sclerosing cholangitis are at higher risk for liver disease or cancer of the bile ducts.

DIAGNOSING DIGESTIVE SYSTEM PROBLEMS

When doctors suspect that a patient is suffering from a digestive system disease, they will run a series of tests to help them diagnose the exact problem. In an *upper endoscopy* (en-DAH-scuh-pee), doctors insert a tube through the mouth and into the esophagus and stomach to see how the upper digestive system is functioning. This way they can examine the esophagus, the stomach, and part of the small intestines. This procedure is used to detect esophageal cancer, GERD, and tumors in the digestive system.

In a *biopsy* (BYE-ahp-see), doctors remove and examine a small piece of tissue from a problem area. The procedure, often performed with a long needle, helps doctors diagnose digestive system cancers.

Doctors also use special machines to scan inside the body and examine the liver, gallbladder, or spleen for abnormalities. Doctors use liver scans, for example, to detect cirrhosis, hepatitis, and tumors. During a liver scan, a radioactive chemical called a tracer is injected into the blood stream. Once the liver has had time to absorb the tracer, doctors take pictures of the liver and spleen with a special camera. The test helps doctors see how well the liver is functioning.

Doctors can also diagnose certain diseases by using blood tests to check for the presence or absence of certain substances in the blood. High calcium levels, for example, may signify esophageal cancer. Other chemicals that doctors look for include bilirubin, albumin, and iron.

An *upper GI* (gastrointestinal) *series* is an X-ray that reveals the esophagus, stomach, and part of the small intestines. Patients drink *barium,* a thick, chalky liquid. The liquid coats the digestive system. Areas coated with barium appear white on the X-rays. The test can reveal abnormalities and problems in the digestive system linings.

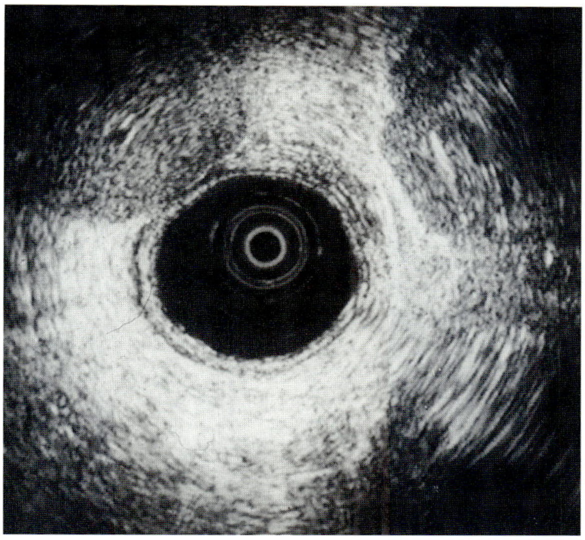

Doctors can examine the inside of the liver and other organs using ultrasound. This ultrasound image shows a healthy liver.

Sometimes called a barium enema, a *lower GI* series uses barium and X-ray images to reveal problems in the large intestine. After the barium has had time to coat the large intestine, X-rays are taken. Doctors use the test to diagnose colon cancer.

Another test to check for digestive abnormalities is the *Bernstein test.* Acid is dripped into the middle esophagus through a tube to see if a person's digestive system problems are the result of acid reflux.

Using *manometry* (muh-NAHM-uh-tree), an inflatable balloon is inserted into parts of the body and then deflated to measure pressure. An esophageal manometry checks pressure in the esophagus and may be used to diagnose GERD. An anal manometry checks muscle function in the rectum and can be used to diagnose Hirschsprung's disease.

In a *colonoscopy* (koh-lun-AH-skuh-pee), a thin tube called an *endoscope* is inserted into the large intestine through the rectum. The tube has a light and a camera that allow the doctor to see inside the colon. Patients are usually given medicine that relaxes them or makes them sleep during the process. A colonoscopy can help doctors detect cancer of the colon or rectum, as well as polyps and other digestive-system disorders.

In *endoscopic* (en-doh-SKAH-pik) *retrograde cholangiopancreatography* (koh-lan-jee-oh-pan-kree-uh-TAH-gruh-fee), or ERCP, doctors insert an endoscope into the mouth, down the throat, and through the esophagus and stomach to examine the liver, gallbladder, bile ducts, and pancreas. Doctors inject dye to help them highlight potential problem areas for X-ray.

An *ultrasound* procedure utilizes sound waves to take pictures of the liver, stomach, and gallbladder. Doctors use the ultrasound to locate gallstones and other digestive system disorders.

In a *gastroscopy* (gas-TRAH-skuh-pee), a thin tube with a light and camera is inserted into the mouth, down the esophagus, and into the stomach to examine the stomach lining for inflammation or damage. This procedure is used to check for gastritis. A biopsy may be performed at the same time.

3

HOW THE ENVIRONMENT AFFECTS THE DIGESTIVE SYSTEM

Like other body systems, the digestive system can suffer from the ill effects of the environment. It is especially vulnerable to the substances we eat or ingest.

DIET AND LIFESTYLE

One way that the digestive system can be affected is by diet and lifestyle choices. Digestive-system disorders caused by an unhealthy diet and poor lifestyle choices include cirrhosis, constipation, and diverticular disease.

CIRRHOSIS

Cirrhosis is a condition in which the liver becomes scarred and damaged and can no longer work properly. Cirrhosis of the liver is a serious and deadly condition. Each year, more than 25,000 Americans die as a result of the disease.

> **Fast Fact**
>
> *Cirrhosis* is a medical term that means "scarring."

One cause of cirrhosis is excessive, long-term abuse of alcohol. Alcoholics usually develop this condition after about a decade of alcohol abuse. Cirrhosis is also commonly caused by hepatitis, especially hepatitis C. (Hepatitis is a swelling of the liver that is often caused by a viral infection.) Other causes of this disease include obesity, inherited and other diseases such as coronary artery disease and diabetes, and some medications and drugs.

Symptoms of cirrhosis include fatigue, nausea, weakness, loss of appetite, and weight loss. People with cirrhosis may also experience other medical problems, including swollen abdomen or legs, easy bruising, bleeding, and jaundice. The disease can lead to other serious medical conditions, including toxins in the blood and brain, bacterial infection, and liver cancer.

While damage to the liver as a result of cirrhosis cannot be repaired, treatment to slow the advance of the disease is available. Treatment of cirrhosis depends upon the cause of the condition. For example, people who have cirrhosis as a result of alcohol abuse should stop drinking alcohol immediately. Other treatments include medicines to battle hepatitis and other cirrhosis-causing conditions. In the most serious cases, doctors may recommend a liver transplant.

CONSTIPATION

Constipation is the most common digestive system complaint in the United States. This condition occurs when the muscles of the large intestine move too slowly, causing stool to absorb too much water as it passes through this part of the body. As a result, a person experiences hard stool that is difficult or painful to pass. People who are constipated may have bowel movements three times a week or less.

> **Fast Fact**
>
> Each year, Americans spend about $725 million on laxatives to treat constipation.

The most common cause of constipation is poor diet—usually a lack of fiber or liquids. Other constipation-causers include lack of exercise, certain medications, and certain medical disorders. Problems with the colon and rectum can also lead to constipation.

To treat constipation, doctors recommend changes in diet to include more fruits, vegetables, and other foods that are high in fiber. *Laxatives,* medicines that soften the stool, may also be prescribed.

Fruits, vegetables, and other high-fiber foods are necessary to maintain a healthy colon and prevent constipation.

DIGESTIVE CANCERS

Diet and lifestyle choices have been shown to contribute to certain digestive-system cancers. Cancer is a disease in which a group of abnormal cells in the body divide without control.

One such digestive system cancer is *esophageal cancer,* or cancer of the esophagus. This type of cancer is a rare but almost always fatal condition. Esophageal cancer usually starts in the inner layers of the esophagus and works its way out. Those most at risk for esophageal cancer are men over the age of fifty.

There are two types of esophageal cancer: squamous cell carcinoma (kar-sih-NOH-muh) and adenocarcinoma (ad-noh-kar-sih-NOH-muh). Smoking, as well as heavy and prolonged alcohol use, increases a person's chances of developing squamous cell carcinoma. Barrett's esophagus, a condition caused by GERD, increases a person's risk for adenocarcinoma.

One of the main symptoms of esophageal cancer is disphagia. *Disphagia* (diss-FAY-jee-uh) is a difficulty swallowing accompanied by the sensation that food has become stuck in the throat or chest. Other symptoms include pain, weight loss, hoarseness, hiccups, and high blood calcium levels.

If the cancer has not spread outside the esophagus, doctors use surgery, followed by chemotherapy and radiation, to treat it. *Chemotherapy* (kee-moh-THERR-uh-pee) is the use of drugs to attack and destroy cancer. According to the National Library of Medicine, about one out of every four patients who has surgery for this cancer will be cured. Once the cancer has spread outside the esophagus, however, it is very difficult to cure. In this case, chemotherapy and radiation may be used to treat a patient's symptoms.

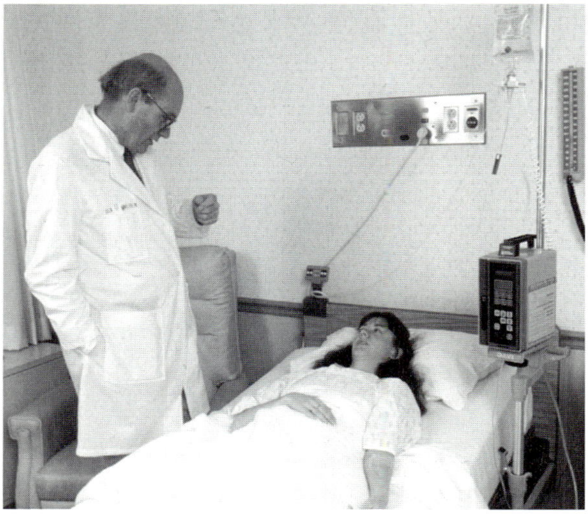

This patient is undergoing chemotherapy treatment for cancer.

Another digestive system cancer is *pancreatic cancer,* a serious and incurable type of cancer that affects the pancreas. According to the American Cancer Society (ACS), pancreatic cancer is the fourth leading cause of cancer deaths in both men and women. Although doctors are unsure what causes most cases of pancreatic cancer, they do know that smoking doubles or even triples the risks of getting this type of cancer. Additionally, African American males over the age of sixty are more likely to be diagnosed with pancreatic cancer.

At first, most patients with pancreatic cancer will not have any symptoms to alert them to the disease. Later symptoms include abdominal pain, weight loss, fatigue, jaundice, vomiting, diarrhea, back pain, and indigestion. Doctors diagnose pancreatic cancer after physical examinations and other tests are made. These may include *magnetic resonance imaging* (MRI) tests, *computerized tomography* (CT) scans, or ultrasounds. An MRI uses magnets and radio waves to create an image of the inside of the body. A CT also creates an image of the inside of the body, but it does so using a thin beam of electromagnetic radiation. An *ultrasound* uses high-frequency sound waves to create a picture of the inside of the body.

Pancreatic cancer is difficult to treat and control effectively. Cancerous tumors can be removed from the pancreas in only two out of ten cases. Other patients with the condition are treated with chemotherapy, radiation, or a combination of both. Once pancreatic cancer spreads to other organs, treatment is usually ineffective. In most of these cases, patients live for about a year. Because pain is a common problem for these patients, easing the pain is a serious concern for doctors who treat the condition. They may prescribe pain medication, radiation therapy to shrink the tumor, or surgery to cut certain nerves in order to ease the pain.

DIVERTICULAR DISEASE

Diverticular (dye-ver-TIK-yoo-lur) *disease* is a condition that affects the colon. Although doctors aren't sure, they think that diverticular disease may be caused by eating a diet that is low in fiber. The disease is much more common in industrialized countries like the United States, where diets are generally low in fiber. There are two types of diverticular disease: diverticulosis and diverticulitis.

Diverticulosis (dye-ver-tik-yoo-LOH-siss) occurs when a small pouch in the colon bulges outward. This condition is common in older people. Half of all people between the ages of sixty and eighty have diverticulosis, while nearly all people over the age of eighty are affected by the condition. While many people with diverticulosis do not have any symptoms, some may experience cramps, bloating, and constipation.

Diverticulosis becomes a serious health problem when a colon pouch becomes infected or inflamed. This condition is called *diverticulitis* (dye-ver-tik-yoo-LYE-tiss). Symptoms include pain in the abdomen, fever, chills, nausea, cramping, and constipation. Doctors treat diverticulitis with antibiotics. If not treated early enough, diverticulitis can lead to more serious health conditions, including intestinal tears, blockages, and bleeding.

FATTY LIVER

Fatty liver, also called steatosis (stee-uh-TOH-siss), is a condition in which fat builds up in liver cells. Some fat is normally present in liver cells. Fatty liver occurs when fat makes up 10 percent or more of the liver. Doctors diagnose the problem by taking a biopsy from the liver and examining the tissues under a microscope.

Fatty liver is usually not a serious condition and probably does not cause any damage to the liver. Fatty liver can, however, lead to inflammation of the liver. It most often affects overweight people and those with diabetes. Other factors, such as a poor diet or alcohol abuse, can also lead to fatty liver. Treatment depends upon the underlying cause of the condition. For example, doctors will recommend that an obese person lose weight. If fatty liver is not treated, it can progress to cirrhosis.

GASTRITIS

Gastritis (gas-TRYE-tiss) is an inflammation of the stomach lining. This condition is usually caused by diet or lifestyle choices, including drinking too much alcohol, smoking, eating hot and spicy foods, and overeating. It may also result from food-borne bacterial infections, some types of medicine, and other health conditions.

Symptoms of gastritis include pain in the stomach, nausea, bloating, and vomiting. Patients with this condition usually take *antacids,* medicines that reduce the amount of acid in the stomach and help the stomach lining heal. If the condition is caused by smoking, doctors will recommend that the patient quit. If a bacterial infection is the cause, they may prescribe antibiotics to kill the bacteria. Gastritis can lead to other serious medical conditions. Chronic gastritis, for example, increases the risks of gastric cancer.

GASTROESOPHAGEAL REFLUX DISEASE (GERD)

Gastroesophageal (gas-troh-ih-sahf-ih-JEE-ul) *reflux disease* is a condition in which the valve between the stomach and esophagus, called the lower esophageal sphincter (LES), doesn't close properly. As a result, stomach acids escape into the esophagus. GERD is a common condition. According to ACS, about half of all adult Americans experience GERD at least once a month. One out of every five United States adults experiences GERD symptoms weekly.

Symptoms include a burning sensation in the middle of the chest, often called heartburn or acid indigestion. People may also have a bitter taste in the mouth. The severity of the condition depends upon how weak the valve is. Mild cases of GERD can be treated with antacids. Frequent use of these antacids can cause serious side effects, however, and they should be taken with care, under a doctor's supervision. In some severe cases, however, other medication or surgery may be required. Diet and lifestyle changes can usually ease GERD. People who are prone to this condition should avoid the foods that cause them to experience heartburn.

Frequent GERD can result in serious damage to the esophagus. One condition that sometimes results is called *Barrett's esophagus.* This occurs when abnormal cells grow on the lining of the lower esophagus as a result of frequent exposure to stomach acids. Although Barrett's esophagus may not cause any symptoms, it does put people at higher risk for esophageal cancer. About 700,000 Americans suffer from this condition.

Avoiding Heartburn

Certain kinds of foods can weaken the valve between the stomach and esophagus. These foods include chocolate, citrus fruits, peppermint, fried or fatty foods, and coffee. In addition, smoking, alcohol, obesity, and pregnancy can also increase the risks of GERD. Overweight GERD sufferers should try to lose weight, and smokers should quit.

Medications that can cause GERD include aspirin, ibuprofen, calcium channel-blockers, tetracycline, and anti-inflammatory drugs. People who suffer from the condition after taking one of these drugs should contact a health professional for alternative medications. Finally, stress can cause heartburn. Finding ways to chill out and calm down can ease GERD symptoms.

INDIGESTION

Indigestion, also called dyspepsia (dis-PEP-see-uh), is an uncomfortable or burning feeling in the abdomen. This condition is usually not serious. It most often results from certain eating habits: eating too quickly, eating high-fat foods, or overeating. In other cases, it may be caused by drinking too much alcohol, stress, certain medications, and other health conditions. Smoking, which weakens the lower esophageal valve, can also cause indigestion. Cigarette smoke causes acids to move from the intestines into the stomach and may damage the esophagus itself.

Symptoms of indigestion include burping, nausea, vomiting, and abdominal bloating. Doctors say that the best way to prevent indigestion is to stay away from the things that cause it. To treat indigestion, doctors often recommend antacids or other medicines.

PANCREATITIS

Pancreatitis (pan-kree-uh-TYE-tiss) is an inflammation of the pancreas. If enzymes being produced in the pancreas become active too soon, they begin wearing away the gland. This painful medical condition has many causes. Excessive alcohol intake is one reason people develop the disease. In other cases, it is caused when a gallstone creates a blockage between the pancreas and the small intestine. Certain medications and some medical conditions can also cause the condition. In some cases, pancreatitis is inherited.

There are two types of pancreatitis, acute and chronic. Acute pancreatitis occurs suddenly and without warning. This condition, although serious, usually goes away with treatment. Chronic pancreatitis is long-lasting and potentially dangerous. This type of pancreatitis can cause permanent damage to the pancreas. Both types of the disease can result in such complications as bleeding, infection, and tissue damage. In addition, enzymes and toxins from the damaged pancreas may enter the bloodstream and damage other organs.

Symptoms of pancreatitis include pain in the upper abdomen, bloating, nausea, vomiting, fever, rapid pulse, and dehydration. Doctors treat pancreatitis by recommending lifestyle changes that will ease the symptoms. People with severe cases of pancreatitis are usually hospitalized and given intravenous fluids until their condition has improved.

EATING DISORDERS

An eating disorder is a condition in which a person has a distorted image of his or her body. In some cases, a patient may practice behaviors that severely limit food intake. These behaviors can lead to undernourishment and other nutritional deficiencies. In other cases, people may overeat, causing obesity and other health conditions. Eating disorders often occur with mental health problems, such as depression and anxiety disorders.

Although eating disorders are classified as mental health conditions, they affect the digestive system as well. In the United States, such conditions are becoming more common. According to the National Association of Anorexia Nervosa and Associated Disorders (ANAD), about 6 million Americans suffer from eating disorders. These disorders usually affect young women, but they can affect men, as well. Girls of fourteen and fifteen are most likely to develop eating disorders. The two most common types of eating disorders are anorexia nervosa and bulimia nervosa.

ANOREXIA NERVOSA

Anorexia nervosa (an-uh-REKS-ee-uh ner-VOH-suh) is a disease in which people starve themselves to become thin. Even patients who are dangerously underweight may continue to insist that they are too fat. An estimated 90 to 95 percent of all anorexia sufferers are female.

This young woman suffers from anorexia nervosa, an eating disorder that causes victims to starve themselves.

Anorexia is often characterized by an obsession with eating and food. Other symptoms of the disease include refusal to gain weight, constant fear of being fat, irregular menstrual periods, and excessive exercise. Later symptoms include dizziness, fainting, memory loss, confusion, and the inability to concentrate.

Anorexia is a serious health condition. A person who doesn't eat properly is not getting enough of the important nutrients needed to help the body function correctly. These improper eating habits can lead to serious health complications, including malnutrition, dehydration, kidney failure, heart problems, and osteoporosis. *Osteoporosis* (ah-stee-oh-puh-ROH-siss) is a musculoskeletal (mus-kyoo-loh-SKEL-uh-tul) condition in which the bones become thin and fragile. The most severe cases of anorexia lead to death. Studies suggest that 5 percent of all people suffering from anorexia die as a result of the condition.

Treating a person with anorexia is often difficult for doctors. Most people with this disease do not believe that there is anything wrong with them. However, the earlier that patients are diagnosed, the greater the chance of helping them to recover quickly. Hospitalization may be an important part of the healing process. In the hospital, patients learn better and healthier eating habits and adjust their views of their body and food.

BULIMIA NERVOSA

People suffering from bulimia sometimes purge their systems of food by forcing themselves to vomit.

Bulimia nervosa (bul-EE-mee-uh ner-VOH-suh) is a disease in which a person binges on food and then either purges the food from the body, fasts, or strenuously exercises. *Purging* is vomiting or using medications such as laxatives or diuretics (dye-yoor-ET-iks) to rid the body of food. *Fasting* is not eating for a period of time. Like those suffering from anorexia, bulimics are excessively concerned with the shape and size of their bodies. As with anorexia, most patients suffering from the condition are young women. ANAD estimates that 5 percent of all female college students suffer from this condition.

Symptoms of bulimia include puffy cheeks or glands caused by frequent vomiting, stained or discolored teeth, and withdrawal or depression. Like anorexia, bulimia can lead to serious health problems, including heart disorders, esophageal inflammation, digestive disorders, blood-pressure abnormalities, and erosion of tooth enamel. It can also lead to heart failure and death.

Doctors treat bulimia in much the same way that they treat anorexia. Both the physical and psychological symptoms of the disorder must be treated. Patients need to learn healthier eating habits and find ways to accept their bodies as they are.

BINGE EATING DISORDER

Binge eating disorder is a condition in which people have frequent episodes of uncontrolled, or binge, eating. During these eating binges, people will eat more than is necessary to satisfy their hunger. After each binge, a patient may feel guilty and unhappy.

Binge eating is one of the most recent eating disorders to come to the attention of health professionals. Although the condition is not accepted as a medical problem by all doctors, the NIDDK believes that binge eating affects as many as 4 million people in the United States—2 percent of all Americans. Binge eating, which is also a feature of bulimia nervosa, affects both men and women, although some studies suggest that 60 percent of all sufferers are female. Binge eating is the most common of all eating disorders.

Eating Disorder Danger Signs

Warning signs of an eating disorder include the following:

- Extreme and dramatic weight loss
- Constant expressions of feeling fat, denying hunger, or anxiety about weight gain
- Preoccupation with food, calories, and nutrition
- Irregular or absent menstrual periods
- Wanting to eat alone or eating secretly
- Compulsive eating, whether hungry or not
- Binge eating followed by purging
- Depression or anxiety
- Brittle hair or nails; loss of hair
- Frequent use of the bathroom after eating
- Signs of frequent vomiting, including reddened fingers or puffy cheeks

Most of the people who suffer from this condition are overweight or obese. Doctors are not sure what causes people to become binge eaters, but depression may be a factor. Symptoms of the disorder include eating large amounts of food at one sitting; eating more quickly than normal; eating alone; and feelings of guilt and depression after bingeing.

Like other eating disorders, binge eating can have serious health consequences. It can lead to high blood pressure, high cholesterol levels, heart problems, diabetes, and gallbladder disease. People with this disorder should see a doctor for treatment of the physical and psychological aspects of the condition.

This young woman chose obesity surgery as her only option to lose weight after numerous failed attempts at dieting. The surgery removed part of her lower intestine. She has lost 29 pounds since the operation.

Obesity

Obesity is the condition of having a body mass index (BMI) of 30 or more. BMI is a measure that compares the amount of fat to a person's weight and height. You can calculate your BMI by multiplying your weight in pounds by 704.5, then dividing the result by your height in inches, and divide that result by your height in inches again.

PICA

> **Fast Fact**
>
> The word *pica* is from the Latin term for "magpie," a bird that will eat almost anything.

Pica (PYE-kuh) is a condition that is characterized by eating nonfood substances for a period of at least one month. People who have this condition may eat dirt, paper, paint chips, clay, sand, and other substances. Young children are most likely to have this condition. According to the National Library of Medicine, 10 to 32 percent of children aged one to six exhibit this behavior. Pregnant women may also crave nonfood substances. Some patients with pica may be unconsciously trying to fill a nutritional deficiency.

Doctors treat pica by working with the patient and his or her family to correct the unusual behavior. In some cases, medicines help. Doctors often take blood tests to make sure that the condition is not causing nutritional deficiencies or abnormalities, such as elevated lead or iron levels in the blood.

Do the Media Cause Eating Disorders?

Why do some people develop eating disorders? According to the National Eating Disorders Association, a variety of factors can lead to eating disorders. These factors include parental or peer pressure to be thinner; a history of physical, sexual, or mental abuse; and a history of dieting. According to the Association, Americans spend more than $40 billion on dieting and diet products.

Some health experts have also pointed a finger at the media. Most women shown in magazines, movies, and television shows are thin. This promotes the ideal American body image as skinny. Although medical experts don't believe that these images are a cause for eating disorders, they do contribute to the problem. Health professionals also point to the increasing number of hours that Americans spend in front of the TV or on the computer—hours that could be spent taking part in physical activity.

DISEASE-CAUSING ORGANISMS

Bacteria and other disease-causing organisms can be found everywhere: in the air that we breathe, the water that we drink, and even the food that we eat. These organisms can cause sicknesses with such symptoms as cramps, nausea, vomiting, diarrhea, and dehydration. They can also lead to more serious health conditions, including kidney failure. In some cases, illnesses caused by bacteria, viruses, and other organisms are fatal.

DIARRHEA

Diarrhea is loose watery stools that occur more than three times in one day. The condition is most often caused by food-borne or water-borne bacteria, viruses, and parasites. It can also be caused by intolerance to certain foods. Other causes include such health conditions as irritable bowel syndrome and lactose intolerance, as well as antibiotics and other medicines.

Diarrhea is a common health problem around the world. The most common cause of bacterial diarrhea is *campylobacter* (kam-pih-loh-BAK-ter), a bacterium most often spread through handling or eating raw or undercooked poultry. Each year, these bacteria cause about 10,000 documented cases of diarrhea in the United States. Researchers estimate that more than 2 million people suffer from the bacteria annually. Most cases of the illness are caused by just one strain of these bacteria.

Another common type of diarrhea is nicknamed "traveler's diarrhea." Each year, as many as 50 percent of all international travelers develop this condition. Travelers often develop diarrhea after being exposed to bacteria, viruses, and parasites in food and water in other countries. The most common cause is the *Escherichia coli* (eh-shur-IK-ee-uh KOH-lye), or *E. coli,* bacteria. People visiting developing countries are most at risk for this health condition.

Diarrhea leads to a lack of fluids in the body, which can cause dehydration. This can create serious problems, especially in young children, the elderly, and those with other health conditions. People who suffer from diarrhea should drink plenty of fluids to keep themselves hydrated. Those suffering the condition for more than three days should consult a doctor.

FOOD-BORNE BACTERIAL DISEASES

The Centers for Disease Control (CDC) estimates that 76 million Americans develop food-borne illnesses each year. As many as 5,000 die as a result of these illnesses. Those most at risk include the elderly, infants, children, and people with other health conditions. Most cases of food poisoning are caused by bacteria that are present in raw meat, poultry, seafood, and eggs. Researchers have identified more than 250 food-borne diseases. Diseases caused by bacteria include botulism, campylobacteriosis, *E. coli* infection, listeriosis, salmonellosis, and shigellosis.

People who suspect that they have become ill from eating contaminated food should contact their doctor immediately. With mild cases, the patient can be treated with an increase in fluid intake. In more serious cases, however, the patient may need to be hospitalized.

Botulism: Terror and Treatment

Some nations have found a way to isolate the poison produced by the bacteria that cause botulism. When sprayed in an aerosol form, this toxin can be fatal. Some researchers worry that botulism could be used by terrorists as a biological weapon. It is easy to produce and could cause serious damage. A 2001 study by Johns Hopkins University in Maryland reported that terrorists could contaminate the United States food supply with the toxin sprayed from airplanes. So far, no nation has used the botulism toxin to do harm.

Botulism can also be used to help people. A protein produced by *C. botulinum* can be used to treat people with eye muscle disease, cerebral palsy, the autoimmune disease multiple sclerosis (skluh-ROH-siss), and migraine headaches. The protein, commonly known as *botox,* is also used by cosmetic surgeons to erase deep lines or wrinkles in a person's face.

BOTULISM

Botulism (BOT-choo-liz-um) is a disease caused by a toxin, or poison, that is released by a bacterium called *Clostridium botulinum* (klos-TRID-ee-um bot-choo-LYE-num). Although rare, botulism is often fatal if not diagnosed and treated quickly. Many cases of this illness are caused by home-canned vegetables, especially asparagus, green beans, beets, and corn. The bacteria have also been spread through chili peppers, tomatoes, and aluminum-wrapped baked potatoes that have not been handled properly.

Botulism affects the body's nervous system. Symptoms include double vision, drooping eyelids, slurred speech, difficulty swallowing, and muscle weakness. If the disease is diagnosed quickly, doctors can give patients an antitoxin to neutralize the poisons in the system. Those with serious cases may need to be hospitalized and put on a ventilator to help them breathe. If left untreated, botulism can lead to paralysis (puh-RAL-ih-siss), or loss of movement, and death.

CAMPYLOBACTERIOSIS

Campylobacteriosis (kam-pih-loh-bak-teer-ee-OH-siss) is a disease caused by campylobacter bacteria. These bacteria are most often transferred to people who handle raw poultry or eat undercooked poultry. They can also be transmitted through nonchlorinated water and milk that has not been pasteurized. The bacteria usually live in the feces of poultry and cattle.

There are several different types of the bacteria. The most common one is *C. jejuni* (jeh-JOO-nee). It is the leading cause of bacterial diarrheal illnesses in the United States. Each year, about 24 million Americans suffer from campylobacteriosis caused by *C. jejuni.* Most sufferers are children under the age of five and young adults.

Symptoms of the disease include diarrhea, nausea, vomiting, abdominal cramping and pain, and fever. Although campylobacteriosis is common, it is rarely fatal. Even without any treatment, it usually runs its course in two to five days. In rare cases, however, patients who have had campylobacteriosis may later develop a disease called Guillain-Barré (gee-LAHN buh-RAY) syndrome, a condition that causes temporary paralysis of the limbs and breathing problems. People who develop Guillain-Barré syndrome may need to be treated in a hospital.

E. COLI INFECTION

E. coli is a bacterium found in the intestines of humans and animals. Some types of *E. coli* are harmless. Others, however, can cause sickness and disease.

Harmful *E. coli* bacteria are most commonly transmitted by cattle. However, *E. coli* bacteria have also been found in sprouts, lettuce, unpasteurized milk and apple cider, and contaminated well water. Symptoms include diarrhea, nausea, abdominal cramps, and fatigue. *E. coli* infection can damage the lining of the intestines. The illness usually passes within five to ten days without treatment.

The most common and serious type of *E. coli* in the United States is a strain known as 0157:H7, or *enterohemorrhagic* (en-ter-oh-hem-or-AJ-ik) *E. coli* (EHEC). Symptoms of EHEC infection include bloody diarrhea and severe cramps, and can cause kidney failure in people with undeveloped or weakened immune systems. Each year, this bacterium sickens about 73,000 Americans and kills more than sixty.

Fast Fact

Some people have developed *E. coli* infections after swallowing improperly chlorinated pool water that is infected with the bacteria.

Another harmful type of the bacteria is enterotoxigenic (en-ter-oh-toks-ih-JEN-ik) *E. coli* (ETEC). ETEC releases toxins that affect the intestinal lining, causing diarrhea. It is the most common cause of traveler's diarrhea.

Although incidences of *E. coli* infection have decreased by 21 percent since 2000, the bacteria still cause problems. In the summer of 2002, officials from the CDC determined that an outbreak in the Midwest and Northeast that made thirty-eight people sick was caused by contaminated ground beef. One person died after being affected by *E. coli.* As a result, some meatpacking companies recalled 415,000 pounds (187,200 kilograms) of meat.

LISTERIOSIS

Another disease that may result in illness and death is *listeriosis* (liss-teer-ee-OH-siss). This food-borne illness is caused by a bacterium called *Listeria monocytogenes* (liss-TEER-ee-uh mahn-oh-sye-TAH-jen-eez). The bacteria can be found in uncooked meats and vegetables as well as in such packaged foods as cold cuts, cheeses, and hot dogs. Every year, about 2,500 people develop listeriosis.

Symptoms of listeriosis range from mild nausea to convulsions and death. Those most at risk for the condition include pregnant women, infants, and people with weakened immune systems. Pregnant women account for one-third of all listeriosis cases. These women are at risk of miscarriages.

In 2002, fifty people in the northeastern United States were diagnosed with listeriosis. The CDC suspected that the source of the outbreak was most likely contaminated deli turkey meat. Most of the people suffering from the disease were hospitalized. Seven people died, and three women suffered miscarriages.

SALMONELLOSIS

Caused by the many types of *salmonella* (sal-muh-NELL-uh) bacteria, *salmonellosis* (sal-muh-nell-OH-siss) affects as many as 4 million Americans each year. About 1,000 of these people die as a result of the condition. Salmonellosis is becoming more common in the United States. One outbreak of the disease can affect hundreds of people at a time.

Salmonella is most commonly found in raw poultry, eggs, and beef. It can also be present on unwashed fruit or raw alfalfa sprouts. People who handle reptiles also risk being affected by the bacteria. Those who are infected with *salmonella* can transmit the bacteria to others while preparing food.

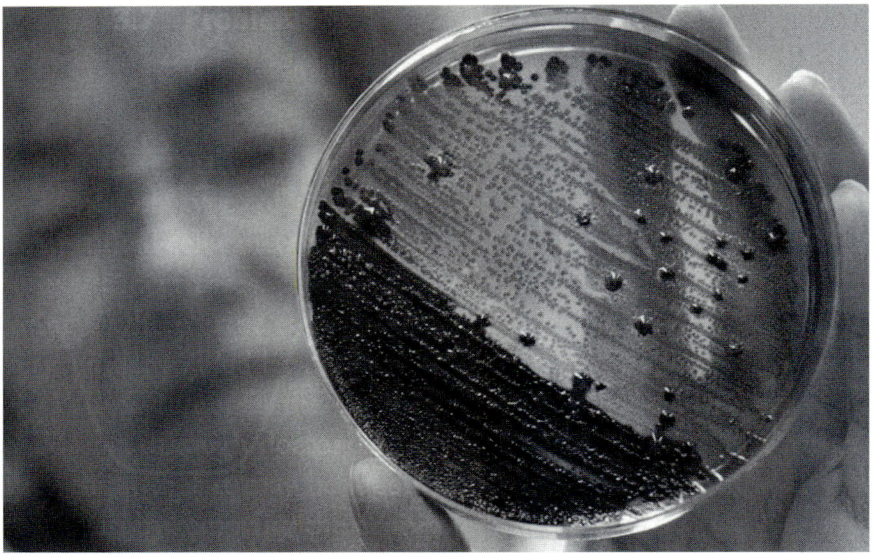

A clinical technologist holds a petri dish with a *salmonella* culture.

Symptoms of salmonellosis include diarrhea, abdominal cramps, fever, nausea, and headaches. These symptoms usually appear between twelve hours and three days after being infected with salmonella. Although the condition usually passes within five to seven days, some people may need to be hospitalized to treat severe diarrhea. In a few cases, salmonellosis can lead to another health condition known as Reiter's syndrome. This condition is characterized by arthritis, painful joints, eye irritation, and pain when urinating. Reiter's syndrome can last for years.

Salmonellosis can become a life-threatening condition if the bacteria enter the bloodstream. As they circulate through the body, the *salmonella* bacteria can attack organs, leading to serious medical conditions and even death. To kill the bacteria, doctors prescribe antibiotics.

Health and History: Scourges of the Past

More than a century ago, food-borne diseases were a common—and often devastating—part of city life in the United States. The most severely affected populations were immigrants and the poor who lived in crowded, unsanitary conditions.

Two of the most deadly diseases were cholera and typhoid fever. *Cholera* (KAHL-er-uh) is an infectious disease that is spread through contaminated food or unclean water. Once contracted, the disease moves swiftly through the body, causing diarrhea, abdominal cramps, vomiting, dehydration, and death—usually within twenty-four hours. Cholera came to the United States for the first time in 1832, when ships filled with immigrants anchored in New York Harbor. In 1847, a cholera epidemic on the East Coast killed 5,000 people in New York City alone.

In the 1800s, *typhoid fever* was another dreaded disease. Caused by *Salmonella typhi,* typhoid fever, like cholera, most often affected people in crowded urban areas. The bacteria can live inside humans for long periods of time. Although some people do not become sick from the bacteria, they can still pass it on to others. These people are called *healthy carriers.*

The most infamous healthy carrier of typhoid fever was an Irish immigrant named Mary Mallon (c. 1870–1938), remembered today as "Typhoid Mary." Mallon arrived in the United States in the early 1900s and worked as a cook. She became the source of a typhoid epidemic in New York City, where she is thought to have infected more than 1,000 people by contaminating their food. Although New York health officials warned Mallon that she carried the deadly disease, she refused to stop working around food. In 1915, the health officials sent Mallon to live on North Brother island in the East River. She never left the island and died there in 1938.

There are more than 2,300 different types of salmonella. One type, Definitive Type 104 (DT104), is resistant to the antibiotics normally used to treat salmonellosis. DT104 is the second most common type of salmonella to affect people. Because it does not respond to traditional treatments, DT104 poses a serious threat to people around the world.

One well-known type of salmonella is *S. typhi* (TYE-fee). Infection by this bacterium can lead to typhoid fever, a serious and often deadly disease that affects millions of people in developing countries each year. This type of salmonella is often transmitted through contaminated drinking water. It is not common in industrial nations like the United States.

SHIGELLOSIS

Also known as bacillary dysentery (BASS-ih-lerr-ee DISS-en-terr-ee), *shigellosis* (shig-uh-LOH-siss) is a health condition caused by *shigella* bacteria. The bacteria are most often transmitted by infected food-service workers who don't wash their hands before preparing food. People can also develop shigellosis after eating vegetables grown in sewage-contaminated fields, drinking contaminated water, or eating food contaminated by flies that carry the bacterium. Shigellosis is most common in tropical and temperate countries, especially in areas that are overcrowded and have poor sewage and water systems.

There are about thirty different types of shigella bacteria. Symptoms of shigellosis include fever, fatigue, diarrhea, nausea, vomiting, and abdominal pain. The disease is usually not serious and passes within five to seven days without treatment. Doctors may prescribe an antibiotic to fight the bacterial infection. A small number of people with shigellosis may later develop Reiter's syndrome.

VIRAL INFECTIONS

Although there are not many viral food-borne diseases, they still pose a serious health problem. Some viruses that can be transmitted through food include the Norwalk virus group, rotavirus, and hepatitis A.

NORWALK VIRUSES

Also known as caliciviruses (kuh-LIH-sih-vye-russ-ez), *Norwalk viruses* are very common sources of food-borne sickness. Each year, about 181,000 Americans suffer from illnesses caused by these viruses. The viruses are transmitted to humans when people eat food that has been prepared by infected food-service workers. In addition, infected fishermen can contaminate oysters as they harvest them. Shellfish can also become contaminated when raw sewage is dumped into the water. In 1993, more than 200 people became ill after eating contaminated oysters from Louisiana.

A person who has been infected with a Norwalk virus usually begins to experience symptoms twenty-four to forty-eight hours after infection. Symptoms include nausea, vomiting, diarrhea, and abdominal pain. The illness lasts for two to three days before disappearing. This type of infection is usually not serious.

ROTAVIRUS INFECTIONS

Rotaviruses are those that can cause severe diarrhea—and even death—in small children. Adults who become infected by rotaviruses often have no symptoms of disease, but when children become infected, serious medical conditions can result. In fact, rotaviruses are the most common causes of diarrhea among infants and young children. According to the CDC, rotavirus infections account for 500,000 visits to doctors each year in the United States. Each year, about 50,000 Americans are hospitalized due to the infections.

Children most often become infected through contact with other children who have the virus. The virus can also be spread through infected food or water. Symptoms of rotavirus infection include diarrhea, nausea, vomiting, and fever. These symptoms can last up to one week. Doctors treat these viral infections by making sure that the patient drinks plenty of fluids and does not become dehydrated. Dehydration, especially in young children, can be deadly.

PARASITIC INFECTIONS

Parasitic infections are diseases caused by parasites in food. There are a number of parasitic infections that cause illness in the United States.

ASCARIASIS

Ascariasis (ass-kuh-RYE-uh-siss) is an infection caused by ascarids, worms that live in the small intestines. Ascarids can grow 6 to 13 inches (15 to 33 centimeters) in length, and some are as thick as a pencil. As many as 100 worms can live inside a single person.

The most common of all human worm diseases, ascariasis usually occurs in developing tropical or subtropical areas and other places with poor sanitation. People often contract the disease through contact with human feces in soil or water. In the United States, ascarid infection is rare.

During the early stages of infection, there are few symptoms of the condition. Later, as more worms develop, a person may experience abdominal pain, difficulty breathing, and blockage of the intestines. Doctors treat this condition with medicines that kill the worms.

CYCLOSPORA INFECTION

Cyclospora (sye-kloh-SPOR-uh) infection is caused by a tiny, one-celled parasite. It is transmitted through food or water that has been contaminated with feces. It infects the small intestines, causing diarrhea, bloating, abdominal cramps and pain, fever, muscle aches, nausea, and vomiting. However, some people who are infected with the parasite do not experience any symptoms at all.

If left untreated, *Cyclospora* infection can last for months. Treatment consists of taking two types of antibiotics to kill the parasites. People with diarrhea should also remember to drink plenty of fluids and get lots of rest.

GIARDIASIS

Giardiasis (jee-ar-DYE-uh-siss) is a condition caused by a tiny, one-celled parasite that is transmitted through water or food that has been contaminated with feces. Once inside the human body, the parasite takes up residence inside the intestines. This hardy, long-lived creature is the main cause of water-borne diarrheal illnesses in the United States.

Symptoms of giardiasis include diarrhea, watery stools, nausea, and stomach cramps. Some people may exhibit no symptoms, however. The condition usually lasts from two to six weeks and can be treated with medicines that kill the parasites.

HOOKWORM INFECTION

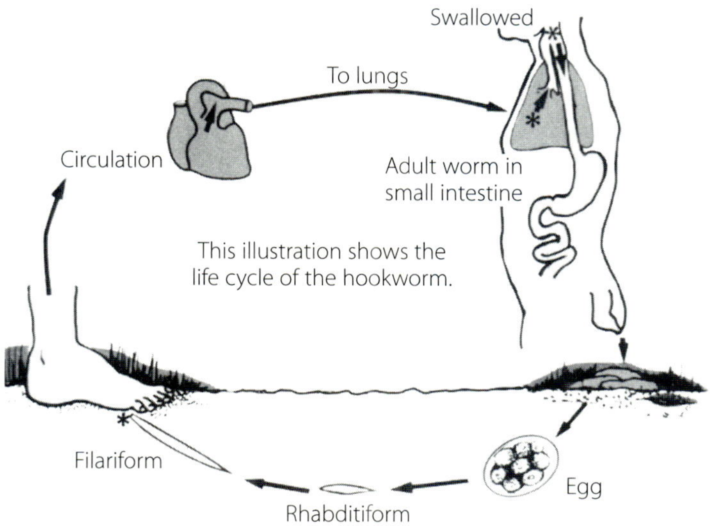

This common infection is caused by tiny parasites called *hookworms.* There are several different types of hookworms, and they are found throughout the world. People most often get the condition when walking barefoot through feces-contaminated soil. Hookworm larvae penetrate the skin of the foot. Then they are carried through the bloodstream to the lungs and from there to the intestines. Symptoms of hookworm infection include diarrhea, nausea, abdominal pain, and cramping. Doctors treat the disease with medicines that kill the hookworms.

PINWORM INFECTION

Another type of infection caused by a parasite is *pinworm infection.* Each year, about 200 million people around the world are infected by these tiny, thread-thin worms. These contagious parasites can survive on bedding, drinking glasses, toilets, food, and eating utensils for up to three weeks. People normally become infected when they touch something that has pinworm eggs on it and then put their hands into their mouths. Children in nursery schools and day-care centers are most at risk.

Once a person has ingested these parasites, they take up residence inside the intestine. Symptoms of pinworm infection include itching in the rectal area, decreased appetite, and weight loss. In some cases, the worms can even be seen around the rectum, and the area may become red and irritated. Doctors treat pinworm infection with medication that kills the parasites.

TRICHINOSIS

Trichinosis (trik-ih-NOH-siss) is caused by a type of *Trichinella* (trik-ih-NELL-uh) parasite. *Trichinella* is most commonly found inside pigs and wild game. The parasite is transmitted when people eat raw or undercooked pork and game. Once inside the human body, the parasite makes its home inside the small intestine, where it lays its eggs. Immature *Trichinella* worms later move into the muscles, where they grow to maturity.

While the worms are in the intestine, patients may experience headaches, fever, chills, coughing, itchy skin, and diarrhea. Once the worms begin moving through the blood vessels and into the muscles, patients may suffer more severe symptoms, including muscle pain and weakness, difficulty breathing, heart problems, coordination problems, and even death. The symptoms usually last for a period of months. Doctors use medicines to kill *Trichinella* in the intestine. Once the infection has moved into the muscles, however, there is no effective treatment. Thanks to federal regulations on animal feed and meat, trichinosis is rare in the United States. It is, however, common in other parts of the world.

Mad Cow Disease

Bovine spongiform encephalopathy (BOH-vyne SPUN-jih-form en-seff-uh-LAH-puth-ee), or BSE, is a neurological disorder that affects cattle. More commonly known as "mad cow disease," BSE is strongly linked to a medical condition in people known as variant Creutzfeldt-Jakob (KROYTZ-felt YA-kobe) disease (vCJD). This illness can cause nervous system disorders, dementia, and death.

The diseases came to public attention in 1986, when cattle in the United Kingdom were diagnosed with BSE. Researchers believe that the sickness was the result of cattle being fed meat products from infected animals. The problem was at its worst in 1993, when 1,000 new cases of BSE were reported in the United Kingdom each week. Whole herds of cattle were deliberately killed, and the United States placed strict regulations on any cows or beef imported from countries known to have infected animals.

The disease took a serious toll on the cattle industry in Europe. In October 2002, researchers in the United Kingdom reported that 2 million cows had been infected. It took a toll on Europe's human population, as well. About 100 people in the United Kingdom and other parts of Europe died as a result of vCJD, contracted from eating contaminated beef.

So far, scientists have been unable to determine exactly what causes BSE. The condition may result from an abnormal form of prion, a protein found on the surface of cells. As of November 2002, just one case of vCJD had been reported in the United States. The patient was a British-born woman who had spent her childhood in Great Britain.

PROBLEM FOODS

Some foods can cause health problems. These foods include shellfish, mushrooms, and red kidney beans.

SHELLFISH POISONING

Shellfish, like those pictured here, can carry the Norwalk virus.

Sometimes toxins from tiny, one-celled sea creatures called *dinoflagellates* (dye-noh-FLAJ-uh-litz) can build up in shellfish and cause illness in people who eat the shellfish. Contaminated mussels, clams, scallops, oysters, and lobsters have all been shown to cause this condition. Depending upon the type of toxins in the shellfish, the contaminated food can cause paralysis, diarrhea, memory loss, and other neurological problems. Symptoms of shellfish poisoning often include diarrhea, nausea, vomiting, abdominal pain, fever, and headache. More serious effects include paralysis, difficulty breathing, seizures, and coma.

If a person suspects that he or she is suffering from shellfish poisoning, the National Library of Medicine recommends calling 911. The condition may require emergency medical treatment. Only a small number of people die as a result of shellfish poisoning.

SCOMBROID POISONING

Scombroid poisoning, sometimes called histamine (HISS-tuh-meen) poisoning, is one of the most common types of fish poisoning in the nation. The illness is caused when certain types of fish begin to spoil. Spoiled tuna and mackerel are the most common causes of scombroid poisoning. As bacteria break down the fish, they create chemicals that are toxic in large quantities.

Symptoms of scombroid poisoning usually appear no more than two hours after a person has eaten the contaminated fish. These symptoms include nausea, vomiting, diarrhea, rash, sweating, and headache. Less common symptoms include burning or swelling of the mouth, accompanied by a metallic taste. This condition usually resolves itself without treatment, although doctors may prescribe antihistamines (an-tee-HISS-tuh-meenz) to quicken the recovery.

CIGUATERA POISONING

Ciguatera (sig-wuh-TER-uh) *poisoning* is caused by eating tropical reef fish that contain an excess of ciguatoxins (SIG-wuh-toks-inz), substances produced by dinoflagellates. Fish most likely to be contaminated with these substances include barracuda, grouper, sea bass, and snapper. Symptoms of the condition usually appear a few minutes to a half-hour after eating the contaminated fish.

Although seldom fatal, ciguatera poisoning is an unpleasant experience. Symptoms include nausea, vomiting, diarrhea, abdominal cramps, sweating, headaches, and muscle aches. Secondary symptoms include a "pins and needles" sensation, the reversal of the sensations of hot and cold, itching, weakness, dizziness, nightmares, and hallucinations. Ciguatera poisoning, which is usually not fatal, runs its course in one to six weeks.

MUSHROOM POISONING

While many people consider wild mushrooms a delicacy, some species of the fungi can cause serious damage to the human body. It is difficult for most people to tell which wild mushrooms are edible and which are poisonous. For this reason, medical experts recommend not eating any mushrooms found growing in the wild.

Symptoms of mushroom poisoning vary, depending upon the type of mushroom and how much was eaten. Symptoms range from nausea, vomiting, abdominal cramping, and diarrhea to organ failure, seizures, and coma. If not treated quickly, patients suffering from mushroom poisoning can die. People who suspect that they are suffering from this condition should seek emergency medical treatment immediately.

Fast Fact

The nation's food supply is getting safer. The incidences of food-borne illnesses have decreased since the mid-1990s.

RED KIDNEY BEAN POISONING

Red kidney beans, when eaten raw or undercooked, can cause sickness. Eating even four improperly cooked beans can cause nausea, vomiting, and abdominal cramping. Although the nausea and vomiting may be severe, symptoms usually disappear within five hours. Other types of beans, including white kidney beans and broad beans, carry the same toxin that causes this illness. However, red kidney beans have the highest concentration of the toxin. Although experts are unsure of the reason, this type of food poisoning is most common in the United Kingdom. No major outbreaks have ever occurred in the United States.

OTHER BACTERIAL AND VIRAL CONDITIONS

Not all bacterial and viral conditions that affect the digestive system are caused by eating contaminated food. Here are a few digestive system disorders that are not food-borne illnesses:

PEPTIC ULCERS

A *peptic ulcer* is a sore on the stomach lining or the duodenum. Nine out of every ten ulcers are caused by a type of bacteria called *Helicobacter pylori* (HEE-lik-oh-bak-ter pye-LOR-rye), which attacks the stomach lining. However, the ulcers can also be brought on by the long-term use of some medicines, as well as by cancerous tumors in the stomach or pancreas. Smoking raises the risk of people developing serious ulcers. Ten percent of Americans will develop a peptic ulcer at some time in their lives.

> **Fast Fact**
>
> Peptic ulcers are not caused by hot, spicy foods or stress.

One of the chief symptoms of a peptic ulcer is abdominal discomfort that occurs two to three hours after a meal or in the middle of the night. This discomfort can usually be relieved with food and antacids. Other symptoms of a peptic ulcer include weight loss, bloating, nausea, burping, and vomiting. Doctors treat this condition with antibiotics to kill bacteria, as well as antacids and other medicines that help the stomach lining heal. People with peptic ulcers can help themselves by not smoking, avoiding aspirin and other medicines that irritate the stomach, and avoiding food products that contain caffeine.

PERITONITIS

The legendary magician Harry Houdini performs a rope escape and a card trick above. Houdini died of peritonitis caused by a blow to the abdomen.

Peritonitis (perr-ih-toh-NYE-tiss) is an infection of the peritoneum (perr-ih-toh-NEE-um), the inner lining of the abdomen. Peritonitis occurs when germs from an infected colon leak into the abdomen. Symptoms of the disease include pain, swelling, and tenderness of the abdomen, fever, nausea, vomiting, and constipation.

Peritonitis is a life-threatening condition that must be treated immediately. Doctors treat the condition by surgically cleaning the abdominal cavity and removing the infected part of the colon. They will also prescribe antibiotics to kill infection-causing organisms.

Health and History: Peritonitis—A Deadly Infection

Before the invention of antibiotics to kill bacteria, people who suffered from infections were often doomed. The infection peritonitis was a common cause of death in earlier times. Some famous people who died as a result of peritonitis include the following:

- Brigham Young (1801–1877), leader of the Church of Jesus Christ of Latter-day Saints, supervised the Mormon move to Utah. In July 1847, Young founded Salt Lake City. He later served as the first governor of the Territory of Utah. Young died from peritonitis caused by a ruptured appendix.
- Hermann Rorschach (1884–1922) was a Swiss psychologist who invented the inkblot test to analyze people's personalities. Shortly after he finished creating the tests, Rorschach was hospitalized because of severe abdominal pains. He died days later of peritonitis, possibly caused by a ruptured appendix.
- Italian-born actor Rudolph Valentino (1895–1926) was the Brad Pitt of his day. His dark good looks and romantic leading roles in silent movies brought him fame and fortune. At the premier of one of his movies, Valentino suddenly became ill. The thirty-one-year-old actor died soon after from peritonitis caused by a perforated ulcer.
- Born Erich Weiss in Budapest, Hungary, Harry Houdini (1874–1926) immigrated to the United States when he was an infant. As an adult, he became an accomplished magician, capturing the imagination of Americans old and young with his amazing escapes from chains, handcuffs, and water-filled coffins. Houdini's career came to a tragic end after he was unexpectedly punched in the stomach by a fan who wanted to test the magician's legendary strength. He died days later of peritonitis.
- James Joyce (1882–1941), one of Ireland's most famous authors, created such literary masterpieces as *Dubliners* and *Ulysses.* He died in Zurich, Switzerland, after an unsuccessful operation to treat a case of peritonitis that had been caused by an ulcer.

The author James Joyce died as a result of peritonitis.

VIRAL HEPATITIS

Hepatitis (hep-uh-TYE-tiss) is an inflammation of the liver. When the swelling is caused by a virus, the condition is known as *viral hepatitis.* Viral hepatitis is a serious medical condition that can permanently damage the liver. Symptoms of viral hepatitis include jaundice, fever, fatigue, nausea, vomiting, and diarrhea.

There are many different types of hepatitis. The three most common types of viral hepatitis are hepatitis A (HAV), hepatitis B (HBV), and hepatitis C (HCV).

HAV is transmitted through food or water that has been contaminated with human feces. Each year, about 27,000 Americans are infected with the disease. HAV can spread when infected food-service workers handle and contaminate food products.

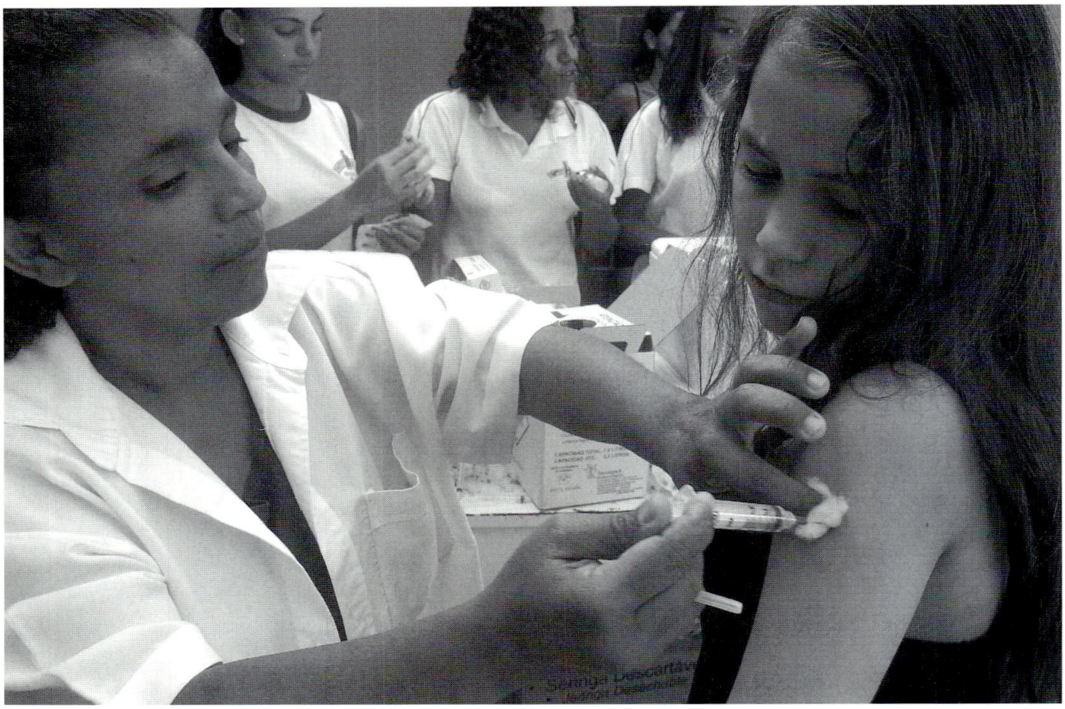

This young woman receives a hepatitis B shot at a health clinic.

HBV is transmitted through contact with a contaminated person's body fluids. It is usually spread through unprotected sexual contact or sharing contaminated needles while using drugs. In recent years, HBV infection has become much more common. According to the American Liver Foundation (ALF), one person out of twenty will be infected with this contagious virus during his or her lifetime. Although most people recover easily from HBV with treatment, a small number suffer permanent liver damage.

Like HBV, HCV is mainly spread through contact with contaminated blood. People who inject illegal drugs, using shared dirty needles most often suffer from this condition. Having unprotected sex with a person who has HCV is also dangerous. HCV is much more serious than HBV. Each year, as many as 10,000 Americans die as a result of this disease. Researchers estimate that this number will triple in the next ten to twenty years.

HCV can remain in the body for many years without causing any symptoms. As time goes on, however, the condition can cause liver damage, including cirrhosis and cancer. Even with treatment, HCV is difficult to get rid of.

Other types of viral hepatitis include hepatitis D, E, and X. Hepatitis X is a form of the disease caused by an unknown virus. Researchers are working to identify this virus. There are vaccines that prevent some types of hepatitis. Currently, hepatitis A and B can both be prevented with vaccines.

LIVER CANCER

Hepatitis B and C can lead to another serious medical condition: liver cancer. *Liver cancer* is any cancer that begins in the liver. While the number of cases of many other types of cancer have decreased in recent years, liver cancer cases have increased. Hepatitis is not the only cause of liver cancer. Other possible *carcinogens* (kar-SIN-uh-jenz), or cancer-causing agents, include the toxic substances produced by some molds, arsenic, and other chemicals. Men over the age of sixty are more likely to develop liver cancer than are younger people.

Liver cancer is sometimes known as "the silent disease." When it first develops, the condition causes few symptoms and may go undetected. Because many cases of liver cancer are discovered after the cancer has grown and spread to other parts of the body, the chances for recovering from this type of cancer are quite low. In later stages, symptoms of the disease include loss of appetite, loss of weight, abdominal pain, nausea and vomiting, and jaundice.

Doctors treat early liver cancer by surgically removing the cancerous tumor or part of the liver. In some case, doctors can replace the liver with a healthy donor liver. When surgery cannot be performed, doctors may use one of the following procedures:

- *Radiofrequency ablation:* Electrodes inserted through the skin are used to kill cancerous cells with heat.
- *Percutaneous ethanol injection:* Alcohol (ethanol) is injected into the tumor to kill cancerous cells.
- *Cryosurgery:* Doctors make a small incision in the abdomen and then use a special instrument to freeze and kill cancerous cells in the liver.
- *Hepatic* (heh-PAT-ik) *arterial transfusion:* A tube called a *catheter* is inserted into the vein that carries blood into the liver. Then cancer-killing drugs are injected into the catheter.
- *Chemoembolization* (kee-moh-em-boh-lih-ZAY-shun): Doctors surgically block off the blood supply to the cancerous tumor. Then cancer-killing drugs are injected into the tumor.

Doctors may also use chemotherapy and radiation treatment to shrink large tumors and make them more compatible with these treatments. Chemotherapy and radiation are also used in late-stage cancer, to make the patient more comfortable and prolong life as much as possible.

4

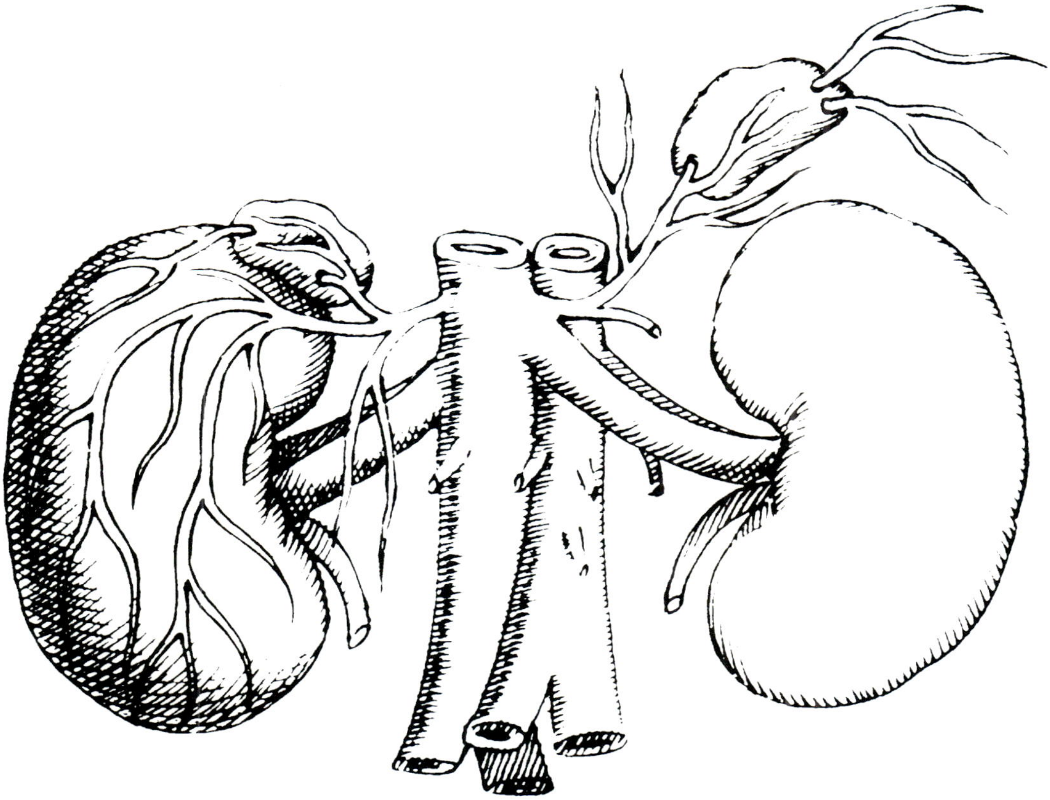

THE URINARY SYSTEM

Without clean blood, the human body could not function properly. The body system responsible for keeping blood free of waste is the *urinary* (YOOR-ih-neh-ree) *system.* The urinary system filters excess water and liquid waste, called *urea,* from the bloodstream and expels it from the body. Urea is a type of waste that is left behind after the body breaks down proteins in food. The urinary system ensures that urea doesn't build up in the bloodstream and cause serious medical problems.

The urinary system is made up of two kidneys, two ureters, the bladder, two sphincter muscles, and the urethra. The urinary system of males and females are nearly identical, except for the length of the urethra. In males, the urethra is longer.

THE KIDNEYS

Fast Fact

Renal is a word that is used to describe anything concerning the kidneys. It comes from the Latin word for kidney, *renes.*

The kidneys, located near the center of the back below the rib cage, are an important part of the urinary system. These two fist-sized, bean-shaped organs perform a number of useful functions. Their most important job is to filter the blood to remove impurities and excess water.

The kidneys contain millions of tiny units called *nephrons.* Inside each nephron is a *glomerulus* (gloh-MER-yoo-luss), a tiny blood vessel that performs the job of filtering waste from the bloodstream, as well as a tube-shaped structure called a *tubule* that collects urine. Anything that the body might use for nourishment—sodium, potassium, or phosphorous—is returned to the bloodstream.

The kidneys perform other important functions. They ensure that the body maintains a proper balance of water and chemicals. If the body already has enough sodium, for example, the kidneys will remove the excess sodium from the body as waste.

Another function of the kidneys is to release three key hormones into the bloodstream. These hormones are erythropoietin (uh-rith-roh-POY-uh-tin), or EPO; renin; and vitamin D. EPO stimulates red blood cell production in the bone marrow, while renin helps control blood pressure. Vitamin D helps the body absorb and maintain calcium, a substance that is important for healthy bones and teeth.

THE BLADDER AND BEYOND

Waste matter, called *urine,* travels from the kidneys to the bladder via the *ureters* (YOOR-ee-terz). Ureters are two long, thin tubes. Each ureter measures from 8 to 10 inches (20 to 25 centimeters) in length and is lined with muscles. These muscles contract and relax, controlling the amount of urine into the bladder.

Urine is stored in the balloon-shaped bladder until a person is ready to urinate. The bladder is a muscular organ, capable of holding up to 2 cups (0.48 liters) of urine for as long as five hours. Sphincter muscles keep the urine from leaking out before its time. Humans know when the bladder is filled because of nerves located there. As the bladder becomes fuller, the urge to urinate becomes stronger.

Humans normally urinate three or four times each day. During urination, the sphincter muscles relax, allowing the bladder to empty. The urine passes from the bladder out of the body via a tube known as the *urethra* (yoor-EE-thruh).

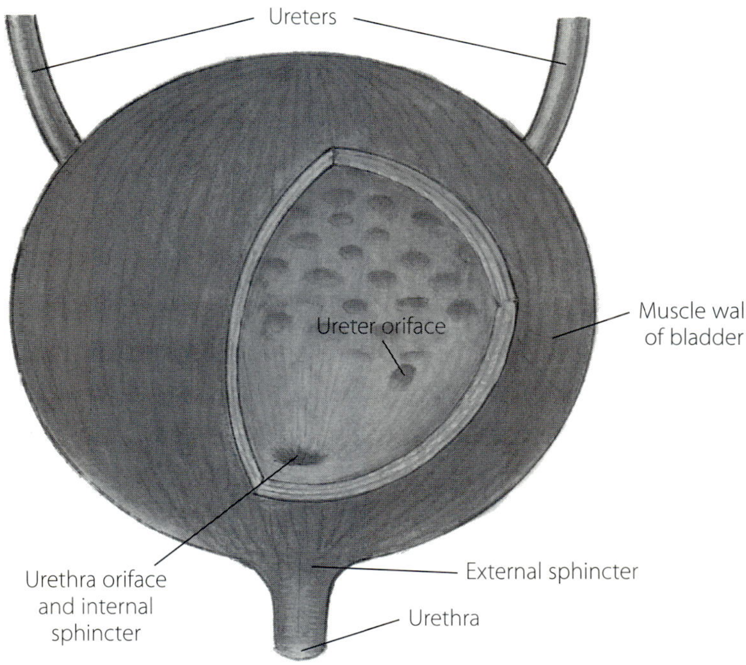

This illustration shows the parts of the bladder.

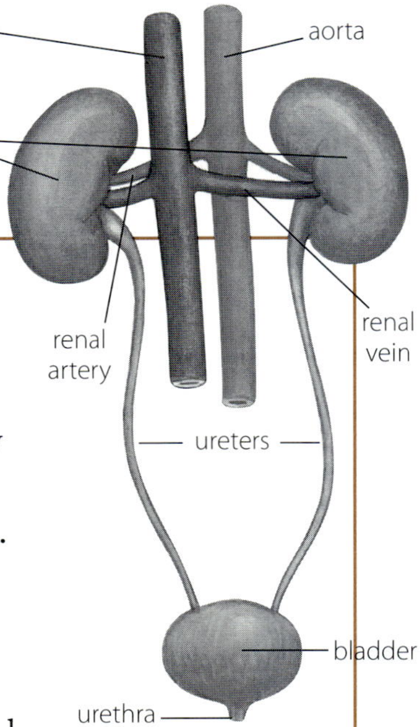

Fascinating Urinary Facts

- Every day, the kidneys process about 200 quarts (190 liters) of blood. They remove about 2 quarts (1.9 liters) of waste and extra water from the blood.
- Each day, humans excrete about 2.5 quarts (2.4 liters) of urine.
- Urine is passed from the ureters into the bladder every ten to fifteen seconds.
- Doctors can use a person's urine to detect medical problems. For example, the presence of sugar in the urine might indicate diabetes. Blood in the urine might indicate an injury or infection.
- The right kidney, located below the liver, is slightly lower than the left kidney.

5

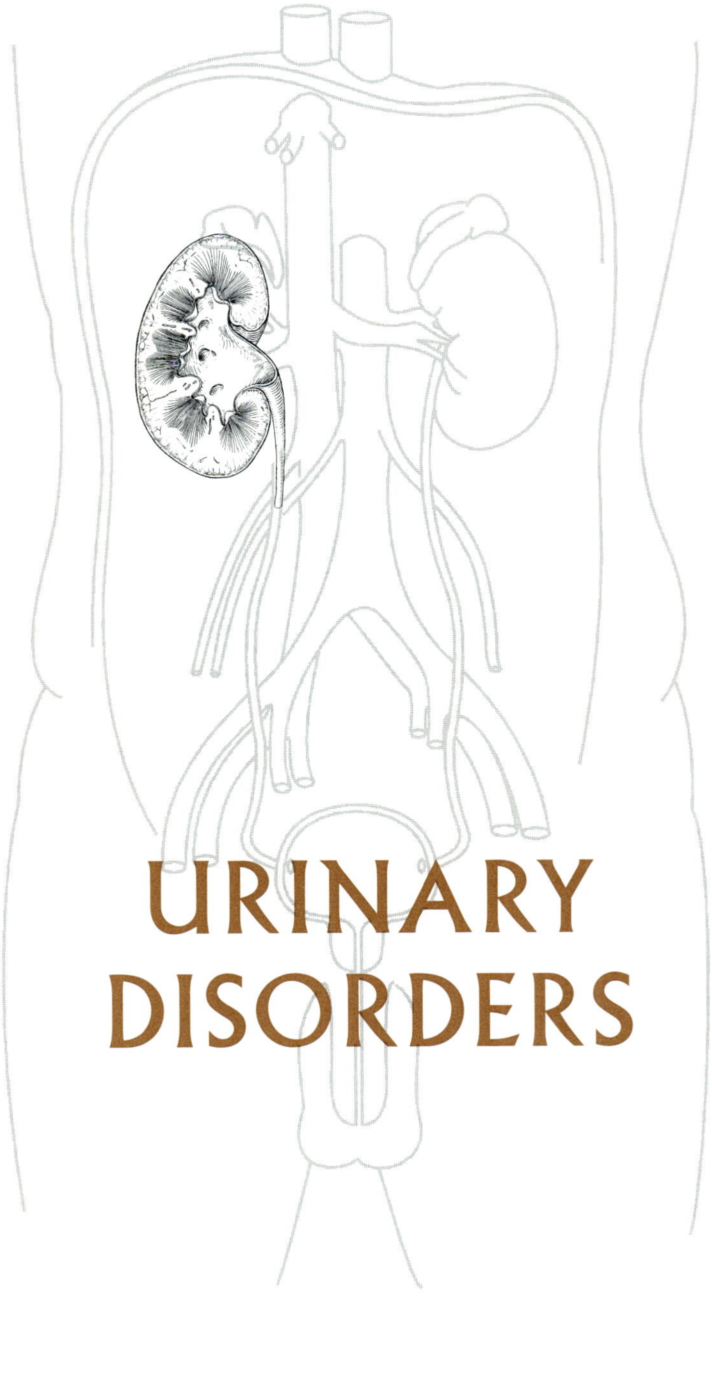

URINARY DISORDERS

Many medical conditions affect the urinary system. These conditions can affect any part of the system, from the kidneys to the bladder. Health problems may result from injury, disease, and even aging. The following are just a few disorders of the urinary system.

Derrick Nix (left), a running back for Southern Mississippi University, was unable to play football with his team as a result of kidney disease.

KIDNEY CONDITIONS

Kidney disease is a serious problem in the United States. Each year, more than 35,000 Americans die as a result of kidney problems. Most kidney diseases attack the nephrons. Symptoms of kidney disease include high blood pressure, swelling of the feet and hands, blood in the urine, frequent urination, and back or side pain.

Diabetes and high blood pressure are the most common causes of severe kidney damage. These conditions can do serious damage to the kidneys over a number of years. This is called "silent damage." Other conditions are congenital in nature, meaning that they are present at birth. Other causes include trauma and aging. As people get older, the sphincter muscles don't work as well as they once did, causing urine leakage or frequent infections.

Kidney disease and damage can lead to a failure of one or both kidneys. Gradual kidney failure over time is called chronic renal failure. Because this type of kidney failure is often not noticed until the disease has reached a late stage, it can be difficult to treat and reverse. That is why it is important for people to consult a doctor as soon as they notice any sign of kidney problems.

When the kidneys fail suddenly, it is called acute renal failure. This might result from an accident, drugs, or poisons. When the kidneys are not seriously damaged, acute renal failure can be treated, and the failure can be reversed.

SIGNS OF KIDNEY FAILURE

Common signs that kidneys are failing include the following:

- *Uremia* (yoor-EE-mee-uh) is a buildup of waste and water throughout the body. Waste matter slowly poisons the body, causing a person to feel exhausted. The hands and feet may swell up. This condition is life threatening and must be treated immediately.
- *Nephritis* (nef-RYE-tiss) refers to any inflammation of the kidney. It might be caused by an autoimmune disorder, medications, environmental agents, or another urinary condition.
- *Edema* (eh-DEE-muh) is swelling in the tissues caused by a buildup of liquids. This condition most often affects the legs or feet.
- *Proteinuria* (proh-teen-YOOR-ee-uh) is a buildup of proteins, particularly albumin and globulin (GLAHB-yoo-lin), in the urine. This condition can cause kidney damage, as well as damage to the circulatory (SURK-yoo-luh-tor-ee) system.
- *Hematuria* (heem-uh-TOOR-ee-uh) is blood in the urine. There are many reasons that a person might have blood in the urine. Some are harmless reasons; others are not. All bloody urine should be analyzed.
- *Albuminuria* (al-byoo-mih-NOOR-ee-uh) is a buildup of the protein albumin in the urine. This condition can indicate diabetes-related kidney damage.

DIAGNOSING KIDNEY CONDITIONS

Some of the early warning signs of kidney problems are frequent urination, painful urination, blood and/or protein in the urine, and high blood pressure. If a doctor suspects that a person has a kidney condition, there are a number of tests that can be used to determine the problem.

Doctors test blood and urine samples to look for signs of kidney disease, including low red blood cell count, traces of creatinine (kree-AT-ih-neen) or urea nitrogen in the blood, or protein in the urine. Creatinine and urea nitrogen are both waste products that healthy kidneys normally remove from the bloodstream. Their presence in the blood indicates that the kidneys are not working right. The presence of such proteins as albumin in the urine indicates that the kidneys are no longer able to separate protein, a useful material, from waste.

In a *renal biopsy,* a long needle is used to remove a piece of tissue from the kidney. The tissue is then examined under a microscope for any trace of abnormalities. In *renal imaging,* doctors use ultrasound, CT scans, or MRIs to take pictures of the kidneys. These tests can detect abnormal growths on the kidneys or blockages between the kidneys and the bladder.

TREATING KIDNEY CONDITIONS

Doctors who treat urinary system problems are called *urologists* (yoor-AHL-oh-jists). Those who deal specifically with kidney problems are called *nephrologists* (nef-RAHL-oh-jists).

People don't need two kidneys to live. People born with one kidney and those who donate or lose a kidney because of disease or injury can lead healthy, normal lives. In fact, thousands of Americans have just one kidney. With the help of dialysis, people can even live without either kidney.

Renal function is a term doctors use to describe the working of the kidneys. Kidney problems are considered severe when renal function drops below 20 percent and is irreversible. This is called *end-stage renal disease* (ESRD), and it is a medical emergency. Without treatment, a patient suffering from ESRD will die. The two treatments for ESRD are dialysis and kidney transplants.

Dialysis (dye-AL-uh-siss) is a procedure that removes toxic substances and wastes from the body. There are two types of dialysis that are used for people with ESRD: hemodialysis and peritoneal dialysis.

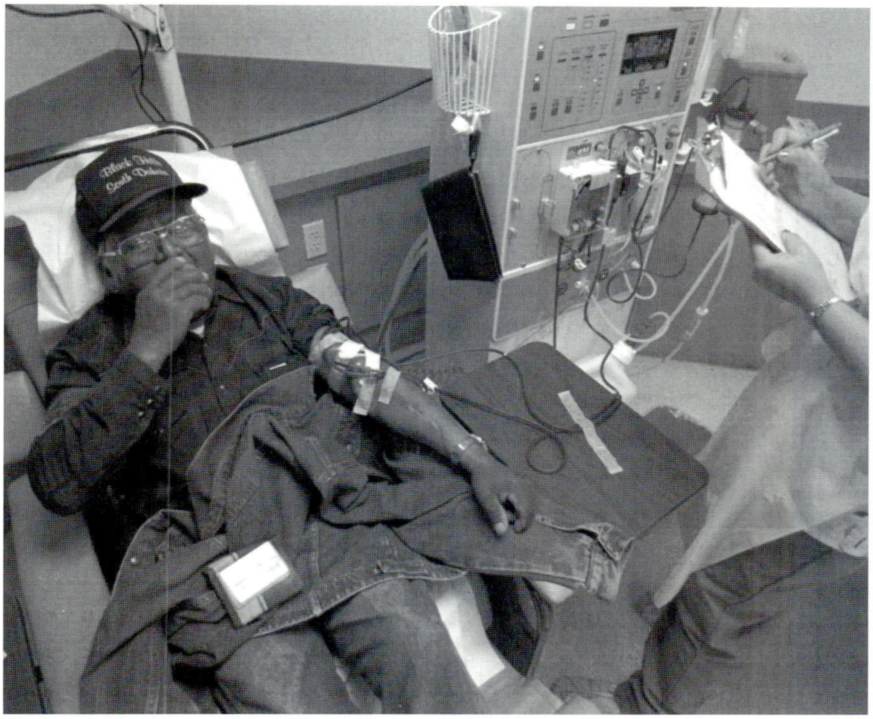

This former uranium miner takes medication while undergoing kidney dialysis in Shiprock, New Mexico. His disease was caused by exposure to radiation while working in uranium mines.

In *hemodialysis* (heem-oh-dye-AL-uh-siss), patients are hooked up to a dialysis machine through a special tube that is inserted into a vein. The dialysis machine removes the patient's blood and runs it through special filters to cleanse it. Once the blood has been cleansed of impurities, it is returned to the body. This procedure, which takes up to five hours to complete, must be performed about three times a week. Most hemodialysis treatments are performed at medical centers, but some can be performed at home.

In *peritoneal* (perr-ih-toh-NEE-ul) *dialysis,* patients give themselves a transfusion of a special solution into the abdomen. This solution filters waste products from the blood. After a few hours, the solution is drained the abdomen. The cycle must be repeated every day.

The most common type of peritoneal dialysis is known as continuous ambulatory peritoneal dialysis (CAPD). Using CAPD, patients perform the dialysis procedure themselves, away from the doctor's office, using a portable device. Another form of peritoneal dialysis can be performed while a patient sleeps at night.

People being treated with dialysis are at higher risk for certain other health conditions. One such condition is *dialysis-related amyloidosis* (am-ih-loy-DOH-siss), or DRA. DRA occurs when a certain protein builds up in the body as a result of dialysis. The condition is common among people who have been on dialysis for five or more years. DRA affects the musculoskeletal system, causing cysts, hollows, and fluid in the bones and joints. Half of all people with DRA also eventually develop carpal tunnel syndrome, a painful swelling of tendons in the wrist. Other conditions that may result from long-term dialysis include polycystic kidney disease (PKD); arteriosclerosis (ar-teer-ee-oh-skluh-ROH-siss), or hardening of the arteries; and osteoporosis.

> **Fast Fact**
>
> Kidney transplants are the second most common type of transplant procedure performed in the United States. The most common type is corneal transplants.

Kidney transplants are surgical procedures in which a patient suffering from ESRD has an unhealthy kidney (or kidneys) removed and receives a donor kidney to replace it. A donor kidney may come from someone who has died recently. However, unlike in other organ transplants, a kidney may also come from a living person who has chosen to donate a kidney. Because a person can function with one healthy kidney, some people choose to give their second kidney to a person in need. Many times, a family member receives the new kidney, but not always. In 1999, more than 13,400 people received new kidneys.

In some cases, a person with severe diabetes who needs a pancreas transplant also has kidney damage as a result of the condition. Such a patient will receive a new kidney and a new pancreas at the same time. A person who has received a new kidney and/or pancreas must take special drugs that prevent the immune system from attacking the new organs.

CONDITIONS THAT CAUSE KIDNEY FAILURE

Diabetes and high blood pressure are the two chief causes of serious kidney damage. In 1997, 42 percent of all cases of ESRD were caused by diabetes, while 25 percent were attributed to high blood pressure. Other major causes of ESRD include several kidney diseases.

DIABETIC NEPHROPATHY

Diabetes is a condition in which the body does not process certain foods, especially carbohydrates, properly. As a result, glucose, a sugar found in carbohydrates, builds up in the bloodstream. Excess glucose causes health problems that can damage the kidneys, heart, and blood vessels.

When excess glucose damages the kidneys, the condition is called *diabetic nephropathy* (nef-RAH-puth-ee). Nephropathy means an abnormal kidney condition. Even when people carefully control their diabetes, diabetic nephropathy can still damage the kidneys. For this reason, diabetics should have their kidneys checked at least once a year.

This young patient undergoes treatment for diabetic nephropathy.

As many as 20 percent of all people with diabetes may have kidney disease. Those most at risk for diabetes include African Americans and Native Americans. These two groups also have a higher risk for kidney damage as a result of diabetes.

Kidney disease as a result of diabetes most often begins in the glomeruli, the parts of the kidney that filter blood. High glucose levels also increase the speed with which blood flows into the kidneys. This further strains the glomeruli and leads to high blood pressure.

During early diabetic nephropathy, most people will not experience any symptoms and will have no idea that their kidneys are not working as well as they should. Later symptoms include nausea, fatigue, jaundice, and swelling of the hands and feet. These symptoms indicate serious kidney failure and require immediate medical attention.

HIGH BLOOD PRESSURE

High blood pressure, also called hypertension, is a medical condition in which the heart has to work extra hard to pump blood through the body. The condition can damage the blood vessels inside the kidneys so that they no longer work effectively. Waste and other matter become trapped in the kidneys. It is important that people whose kidneys have been damaged keep their blood pressure at proper levels to prevent any further damage. Normal blood pressure is 120 over 80. Those with high blood pressure should ask a doctor about medications and lifestyle changes that might lower their blood pressure.

HIV

The *human immunodeficiency* (im-myoon-oh-dih-FISH-en-see) virus (HIV) is the virus that causes the disease *acquired immune deficiency syndrome* (AIDS). HIV infection can lead to kidney damage, especially damage of the glomeruli. As many as 10 percent of all people with HIV will suffer from kidney failure. People who have HIV-related kidney failure suffer a rapid deterioration. Most cases progress to ESRD within a year of being diagnosed. Doctors are currently looking for ways to slow this deterioration.

CONDITIONS THAT RESULT FROM KIDNEY FAILURE

Kidney failure can lead to a number of serious conditions that require medical attention and treatment. Complications of kidney failure include anemia and renal osteodystrophy.

ANEMIA

Anemia (uh-NEE-mee-uh) is a lack of red blood cells in the body. Some unhealthy kidneys no longer produce EPO, the hormone that stimulates red blood cell growth. As a result, some people with kidney disease develop anemia. This condition can result in fatigue, weakness, and heart problems.

As kidney disease becomes worse, so does anemia. Nearly all people with ESRD have anemia. Doctors treat anemia caused by kidney disease by injecting patients with artificial EPO to stimulate red blood cell growth. Patients must also take iron supplements or receive iron injections to boost their blood iron levels.

RENAL OSTEODYSTROPHY

Renal osteodystrophy (ah-stee-oh-DISS-truh-fee) is a bone condition that occurs when failing kidneys can no longer maintain the proper amount of calcium and phosphorous in the blood. Both of these minerals are important to healthy bones.

Renal osteodystrophy affects 90 percent of all patients on dialysis. It is most harmful to children on dialysis, because their bones have not yet finished growing. If growing bones do not get enough calcium, they become soft and weak. Children on dialysis may then develop renal rickets, a medical condition in which their legs bow outward.

Symptoms of renal osteodystrophy include bone and joint aches. Doctors may prescribe calcium supplements or medicines that replace the calcium in the body. Diets that are high in calcium and low in phosphorous are also recommended.

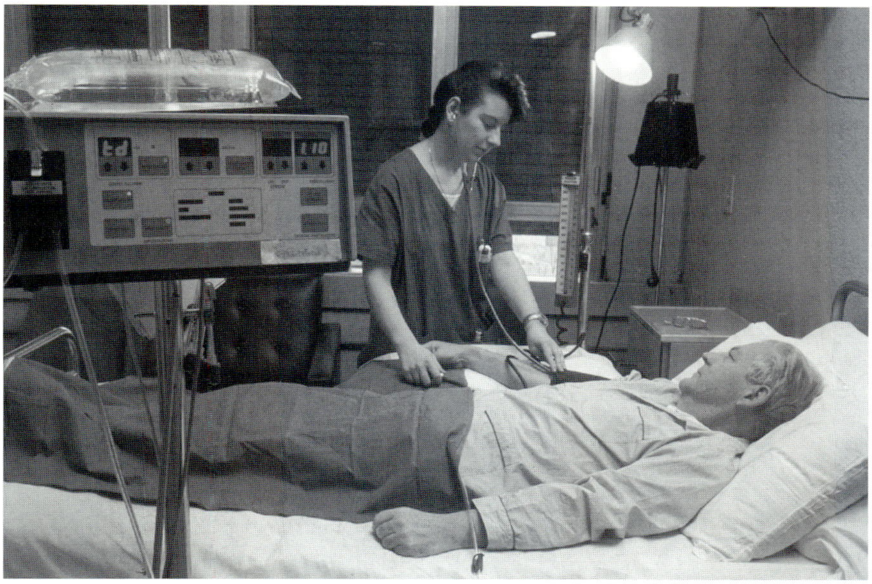

Renal osteodystrophy affects many people who must undergo dialysis.

AUTOIMMUNE CONDITIONS

Some urinary system conditions are autoimmune disorders. These disorders occur when the immune system malfunctions and begins to attack healthy body parts.

GOODPASTURE'S SYNDROME

This rare autoimmune disorder harms both the kidneys and the lungs. It causes inflammation of the glomeruli and can seriously damage the kidneys. Goodpasture's syndrome, which affects mostly young men, can last anywhere from two weeks to two years. Doctors believe that the condition may be triggered by viral infections or chemicals called *hydrocarbons,* contained in some glues and gasoline.

A customer fills his car with gas using a vapor recovery nozzle in Burlington, Vermont. Gas stations throughout the state have installed equipment to reduce air pollutants from gasoline emissions.

Symptoms of Goodpasture's syndrome include bloody or dark urine, bloody saliva, decreased amounts of urine when urinating, difficulty breathing, nausea, and vomiting. Doctors treat the condition with medicines that suppress the immune system, as well as blood-cleaning treatments to remove the antibodies that are causing the kidney damage. However, if the disease is not caught quickly enough, serious damage to the kidneys may be permanent and may require dialysis or a kidney transplant.

IGA NEPHROPATHY

IgA nephropathy, also known as Berger's disease, is an autoimmune disorder in which an antibody called immunoglobulin (im-myoon-oh-GLAHB-yoo-lin) A builds up in the glomeruli of the kidneys. The antibody prevents the glomeruli from properly filtering the blood that enters the kidneys, causing illness and kidney damage. IgA nephropathy most often affects young men under the age of thirty. One in four patients suffering from the condition will eventually develop ESRD.

Doctors believe that there is a genetic factor to IgA nephropathy, because it runs in some families. One symptom of the disorder is dark or bloody urine. There is no cure for the condition. Instead, doctors do what they can to prevent serious kidney damage. This may include prescribing medicines to lower blood pressure and special diets.

LUPUS NEPHRITIS

Lupus nephritis is caused by *systemic lupus erythematosus* (LOOP-us eh-rith-uh-muh-TOH-siss), an autoimmune disorder that most commonly affects the musculoskeletal system. The disease causes damage to the glomeruli and may lead to permanent kidney damage. Symptoms of lupus nephritis include bloody or foamy urine, swelling of the feet and hands, and high blood pressure. Doctors treat the condition with medicines that suppress the immune system, reduce inflammation and swelling, and lower blood pressure. As with other kidney diseases, serious damage to the kidneys needs to be treated with dialysis or a kidney transplant.

BIRTH DEFECTS AND GENETIC DISORDERS

A *birth defect* is any physical problem with which an infant is born. *Genetic disorders* are abnormalities in a person's genes or chromosomes. These defects may be inherited from a parent, or they may develop on their own. There are many different urinary diseases that result from birth defects or genetic abnormalities. These diseases include the following:

ALPORT SYNDROME

The inherited disease *Alport syndrome* affects the glomeruli in the kidneys. In addition, some people with the condition also suffer deafness and eye problems. Alport syndrome most often affects males.

In early stages of the disease, those with Alport syndrome will not experience any symptoms. Later, symptoms include blood in the urine, loss of hearing and vision, coughing, and swelling around the feet, ankles, and eyes. There is no cure for Alport syndrome. Instead, doctors monitor and prevent serious kidney damage.

BLADDER EXSTROPHY

Bladder exstrophy (EKS-truh-fee) is a condition in which a child is born with its bladder inside out and protruding through the abdominal wall. The pelvic bones are also separated. According to the March of Dimes, bladder exstrophy affects one out of every 30,000 infants. Boys are more frequently born with this birth defect than are girls. The condition is usually corrected with a number of surgeries. The first are performed shortly after the infant is born. Other surgeries may follow later, depending upon the severity of the condition. The goal of the surgeries is to restore bladder function to the patient and repair problems with the pelvic bones.

EPISPADIAS

Epispadias (ep-ih-SPAY-dee-uss) is an inherited condition that occurs when an infant is born with his or her urethral opening in the wrong place. Instead of being located at the head of the penis, the opening for male babies may be located on the top or side. With female infants, the opening may be in the abdomen. The condition, which often occurs with bladder exstrophy, is more likely to occur in male infants than in females. The treatment for this condition is usually surgery.

HORSESHOE KIDNEY

This congenital condition occurs when a baby is born with its kidneys fused together at one point. The kidneys may work properly, and about one-third of all patients with the condition never experience any problems as a result. However, horseshoe kidney can create health complications. People born with this condition are at a higher risk for other urinary tract problems, including infections, kidney stones, and swelling of the kidneys. In addition, people with this condition may be at higher risk for renal cancer. About one in 400 babies in the United States is born with this condition. More males than females are born with horseshoe kidney.

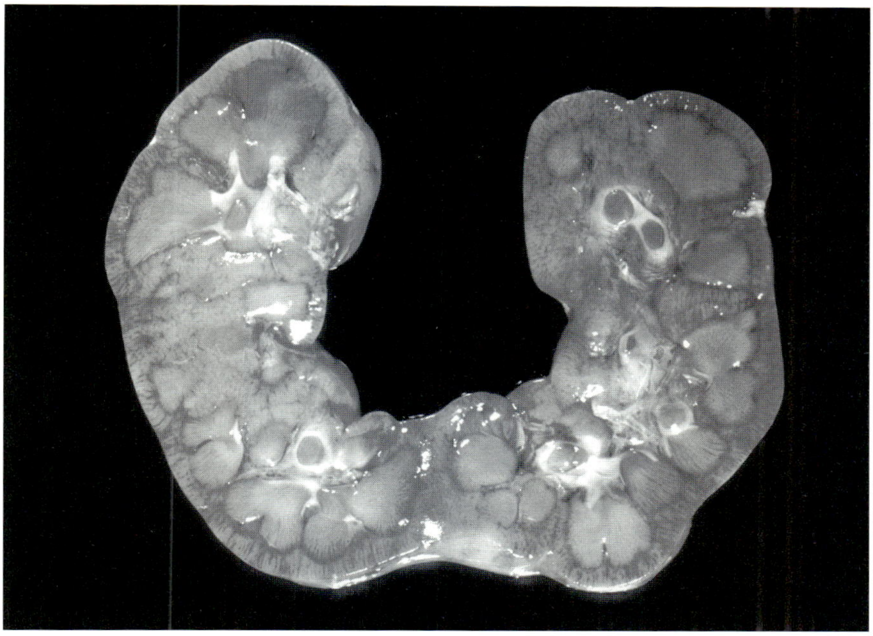

This photo shows a horseshoe-shaped kidney that occurs when two kidneys are fused together.

HYPOSPADIAS

Hypospadias (hye-poh-SPAY-dee-uss) is a condition in which a male infant is born with the opening of the urethra located on the underside of the penis. Surgery must be performed to fix the defect and allow the child to urinate normally. If corrective surgery is not performed, the child will have to urinate sitting down and will have painful sexual intercourse as an adult. This common condition affects about 1 percent of all infant boys.

POLYCYSTIC KIDNEY DISEASE

Polycystic (pahl-ee-SISS-tik) *kidney disease* (PKD) causes multiple, fluid-filled cysts (small, abnormal growths) to form in the kidneys, replacing healthy tissue. This results in decreased kidney function and, eventually, kidney failure. People may have the condition from childhood but not have any problems until their twenties or thirties.

At that time, the patient may begin suffering frequent urinary tract infections (UTIs), blood in the urine, pain, high blood pressure, and kidney failure. Because there is no cure, the disease eventually results in ESRD.

PKD is the fourth leading cause of kidney failure in the United States. About 500,000 Americans currently suffer from this disease—one of the most common types of genetic disorders. The most common form of the disease is called *autosomal* (aw-toh-SOH-mul) *dominant PKD.* This type of PKD usually first affects people between the ages of thirty and forty. A rarer form, *autosomal recessive PKD,* affects infants before they are even born.

Dr. Susan Light is a medical researcher who helped create a kidney transplant anti-rejection drug. Her son Jake, who suffers from kidney disease, will receive one of his mother's kidneys and the new drug she developed.

RENAL AGENESIS

When an infant is born without one or both kidneys, the condition is known as *renal agenesis* (ay-JEN-eh-siss). Those born missing both kidneys cannot survive. This condition occurs in one in 4,000 births. More commonly, children are born missing one kidney. This occurs about once in every 550 births. Many children born with just one kidney can lead normal, healthy lives. They may, however, be at increased risk for such health conditions as kidney infections, kidney stones, kidney failure, and high blood pressure.

VESICOURETERAL REFLUX

Vesicoureteral (vess-ih-koh-yoor-EET-er-ul) reflux (VUR) is a condition in which urine from the bladder flows back into the ureters. In severe cases, urine from the bladder may reach the kidneys, causing scarring and other damage. VUR, which commonly affects children, can be caused by a structural defect in the ureters (primary VUR) or a blockage in some part of the urinary system (secondary VUR). Doctors believe that this may be an inherited condition, because it tends to run in families.

The chief symptom of VUR is a UTI. One out of three children with a UTI is found to have VUR. Other symptoms include bedwetting, high blood pressure, and kidney failure. While most children outgrow VUR, some may need treatment. Treatments include antibiotics to kill infections and surgery to correct any structural defects that may be causing the problem. If VUR remains untreated, a patient may suffer high blood pressure and kidney damage.

MISCELLANEOUS CONDITIONS

Doctors are still not sure what causes some urinary system diseases. Here are a few conditions that medical researchers are still working on.

CYSTOCELE

Cystocele (SISS-toh-seel) occurs when the wall between the bladder and the vagina weakens, allowing the bladder to fall into the vagina. This condition causes urination problems, including bladder leakage when a woman sneezes or coughs and an inability to empty all the urine in the bladder.

Cystocele can be caused by straining or heavy lifting. In some women, the condition occurs when they begin to go through menopause. *Menopause* is the time in women's lives when they normally stop menstruating and their bodies produce fewer of such important hormones as estrogen (ESS-troh-jen). Estrogen helps keep the vaginal muscles strong. For this reason, doctors may recommend that women going through menopause take replacement estrogen.

Doctors treat cystocele in different ways, depending upon how serious the condition is. Mild conditions may not be treated at all. In other cases, doctors suggest a *pessary* (PESS-uh-ree), a plastic device that is inserted into the vagina to hold the bladder in place. Serious cases require surgery to move the bladder back into place.

INTERSTITIAL CYSTITIS

Sometimes called painful bladder syndrome, *interstitial cystitis* (in-ter-STIH-shul siss-TYE-tiss), or IC, is a chronic condition that causes the bladder to become inflamed. Bladder inflammation can result in scarring, stiffening, and bleeding of the bladder, as well as a decrease in the amount of urine that the organ can hold. In some cases, IC can cause ulcers on the bladder. Doctors are not sure what causes IC. It may be an autoimmune disorder that follows a bladder infection, or it may be caused by a bacterial infection.

IC affects more than 700,000 Americans. Nine out of ten sufferers are women. The disease normally first occurs when a person is between the ages of thirty and forty. Symptoms include pain when urinating, frequent urination or feelings of needing to urinate, and pressure or tenderness in the bladder area. People with severe cases may urinate as often as sixty times a day.

There are no specific tests to diagnose IC. Instead, doctors must first rule out other urinary disorders before diagnosing IC. Once the condition has been identified, doctors treat it by trying to relieve any symptoms. One such treatment is *bladder distention.* During this procedure, the bladder is distended, or enlarged, and filled with a liquid or gas. Doctors are not sure why, but the treatment seems to provide some relief for IC sufferers. Another common treatment is a *bladder installation,* also known as a bladder wash or a bladder bath. During this procedure, doctors use a catheter to fill the bladder with a special drug.

Other treatments for IC include pain-relief medications and a healthy diet and lifestyle (no smoking and plenty of exercise). In severe cases, surgery may be performed. Although treatment can provide temporary relief for IC, there is presently no cure for the disease.

RENAL TUBULAR ACIDOSIS

Renal tubular acidosis (RTA) occurs when the kidney does not secrete acids into the urine, causing a buildup of acids in the blood. An excess of acid in the blood can impede growth and cause kidney stones, kidney failure, and bone disease. RTA causes the body to lose potassium, bicarbonate, glucose, calcium, amino acids, and phosphates in the urine. As the levels of these important nutrients drop, patients may experience heart problems and may even die.

While RTA may occur by itself, it is often a complication caused by other medical conditions, especially those that affect the functioning of the kidneys. Conditions that can lead to RTA include lupus, chronic UTIs, kidney-transplant rejection, and an overactive parathyroid (paar-uh-THYE-royd) gland. The parathyroid glands produce a hormone that regulates calcium and phosphorus use in the body.

Symptoms of RTA include fatigue, weakness, confusion, rapid breathing, and bone and muscle problems. Musculoskeletal conditions result because an excess of acid in the body wears away calcium, causing bones to soften and muscles to weaken. It is especially important that children with RTA, whose bones are still growing, be treated.

Doctors treat RTA by prescribing medicines that restore the body's acid levels to normal. Calcium and vitamin D supplements can help prevent bone and muscle damage in the body. With early treatment, most cases of RTA disappear without causing any serious, permanent health problems.

URINARY INCONTINENCE

Urinary incontinence is a condition marked by involuntary urine leakage. This condition occurs when sphincter muscles weaken as a result of infection, medications, or other medical conditions. One out of every ten people over the age of sixty-five is affected by urinary incontinence,

Fast Fact

Many people incorrectly think that urinary incontinence is the normal result of aging. Although it is usually correctable, incontinence is one of the main reasons that elderly people are confined to nursing homes.

though the disease is more common in women. When children experience urinary incontinence, it is often the sign of another medical problem. Health problems that might cause bladder leakage include UTI, spinal injury, neurological problems, or abnormalities of the urinary system.

Two common types of urinary incontinence are stress incontinence and urge incontinence. *Stress incontinence* occurs when a patient's pelvic or sphincter muscles are so weakened that they cannot hold urine in the bladder during such physical activities as exercise, sneezing, or laughing. This type of incontinence often occurs after certain medical procedures or while taking certain medications. Stress incontinence is the most common type of incontinence experienced by women.

Urge incontinence, also known as overactive bladder or irritable bladder, is a condition in which a person feels a strong urge to urinate, followed by the involuntary leakage of a large amount of urine. This type of incontinence often occurs as a result of other medical problems, such as spinal trauma, stroke, multiple sclerosis, and other urinary and bladder problems.

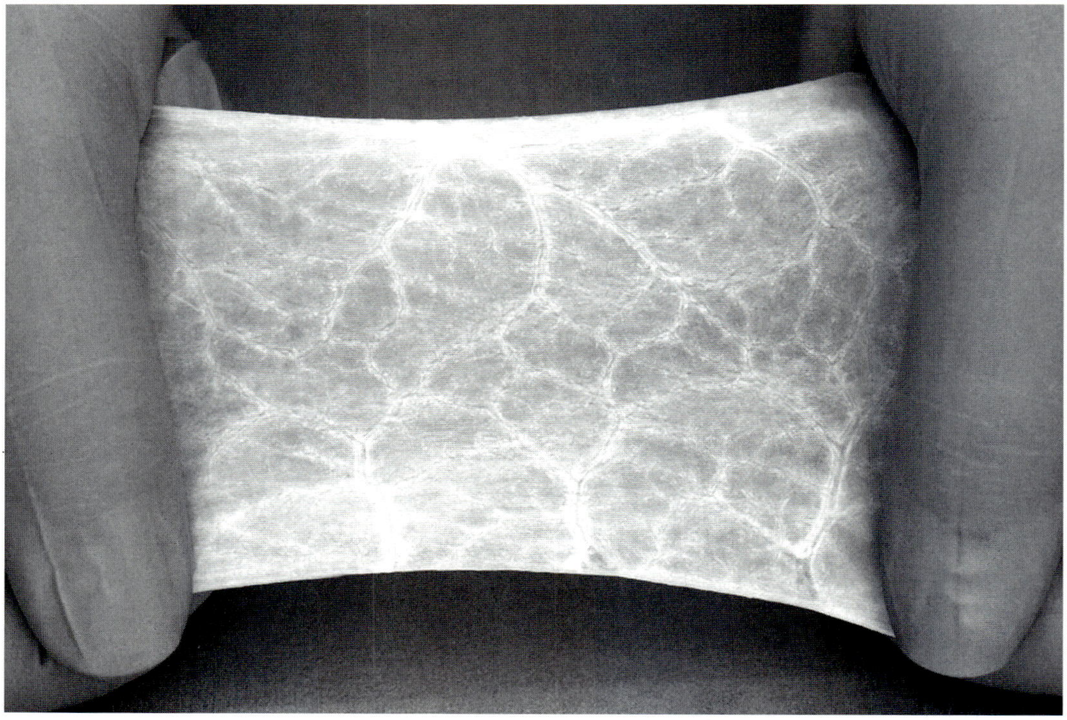

Doctors are using a new material made from pigs' intestines to treat urinary incontinence in women.

Once doctors diagnose what type of incontinence a person is suffering from, they can decide upon a course of treatment. In some cases, urinary incontinence can be treated with exercises that strengthen the pelvic and sphincter muscles. There are also medicines available to take care of the problem. If the leaking is due to a structural problem, surgery may be necessary.

URINARY STONES

Urinary stones are stones that form in a part of the urinary system—in the kidneys, the ureters, or the bladder. The stones often form as a result of excess salts and minerals in the bloodstream. They vary greatly in size, shape, color, and composition. These stones can be as small as a grain of sand or as big as a golf ball. They may be smooth or jagged. Some are made mostly of calcium, while others are made of magnesium, uric acid, or ammonia. The stones often form as a result of excess salts and minerals in the bloodstream, although doctors do not know exactly what causes urinary stones to form.

Urinary stones are a common condition, affecting an estimated 12 percent of all Americans. Those most at risk are white men between the ages of forty-five and fifty-five. Other health conditions, such as UTIs, bowel disease, or an enlarged prostate, increase the risk of getting urinary stones. A person who has one urinary stone is more likely to develop others.

Beverages and Kidney Stones

Recent research has shown a link between recurring kidney stones and cola drinks. A study showed that male kidney-stone sufferers who cut down on their consumption of cola were less likely to develop new kidney stones. Another beverage that seemed to raise the risk of kidney stones was grapefruit juice. Researchers have found that tea, coffee, and wine seem to lower the risks.

Kidney stones are the most common of all urinary stones. Each year, about 1 million Americans are affected by them. Most kidney stones are made up of calcium crystals that separate from urine and collect on the inner walls of the kidney. Although kidney stones originate in the kidney, they can lodge anywhere in the urinary system.

The main symptom of kidney stones and other urinary stones is pain, especially as the stone is expelled from the kidney through the ureter. Other symptoms include nausea, blood in the urine, and feelings of having to urinate.

To pass small kidney stones naturally while urinating, doctors recommend that patients drink lots of water and take medicines to ease pain. When the stones are too large to pass, doctors may use the following treatments:

- *Extracorporeal* (ecks-truh-kor-POR-ee-ul) *shock wave lithotripsy* (LITH-uh-trip-see), or ESWL: In this procedure, doctors use shock waves to break the stones into smaller pieces. Smaller stones can be more easily passed from the body in urine.
- *Percutaneous nephrolithotomy* (per-kyoo-TAY-nee-uss nef-roh-lith-AH-tuh-mee): This procedure, sometimes called tunnel surgery, involves making a small cut in the back and removing the stone.
- *Ureteroscopic* (yoor-ee-ter-oh-SKAH-pik) *stone removal:* During this procedure, doctors insert a special instrument through the bladder and into the ureter. The stone is captured and removed.

Another type of urinary stone is a bladder stone. Bladder stones form when minerals collect inside the bladder. About 95 percent of all people affected by this type of stone are male. Symptoms of bladder stones include feelings of having to urinate, difficulty or inability urinating, bloody or dark urine, and pain in the abdomen or penis. Doctors may use ESWL or another method to remove bladder stones. If not removed, they can cause permanent damage to the bladder or kidneys.

Patients who have a tendency to develop urinary stones may be prescribed medication that controls the condition. In addition, doctors may recommend that these patients drink lots of water every day and avoid foods that are high in calcium and oxalates. Such foods include organ meats (liver, kidneys), chicken, certain types of fish, and chocolate.

DIAGNOSING URINARY SYSTEM CONDITIONS

Problems in other sections of the urinary system beside the kidneys can be diagnosed in a number of ways. One method is a urine test. Also called a urinalysis, a urine test can detect the presence of blood, proteins, and excess minerals in the urine.

For an *intravenous pyelogram* (PYE-uh-loh-gram), doctors inject a dye into the patient's body. The dye collects in urine in the bladder, and X-rays can then be taken to diagnose bladder cancers.

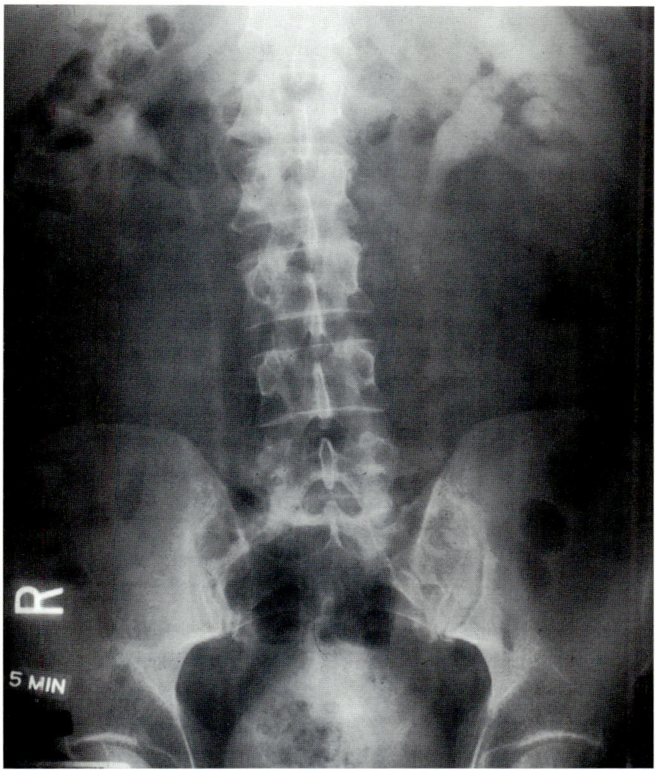

This X-ray shows an intravenous pyelogram of the urinary system.

When performing a *cystoscopy* (siss-TAH-skuh-pee), doctors insert a thin tube with a light and camera through the urethra into the bladder. They can then examine the bladder lining.

A biopsy can be performed during procedures such as a cystoscopy or it can be done surgically. Tissue is removed from a site and examined under a microscope for abnormalities.

Imaging tests that can be used to diagnose urinary disorders include X-rays, CT scans, MRIs, and ultrasounds. All these tests allow doctors to see what is going on inside the body. Each test provides different information, and the doctor will decide which test is best for the situation.

Urodynamic (yoor-oh-dye-NAM-ik) *tests* allow doctors to assess how well the bladder stores urine. Patients are catheterized in which a small, thin tube is inserted through the urethra and into the bladder. Doctors then inject water, gas, or dye into the tube. Imaging tests of the bladder indicate how well that organ is functioning. A urodynamic test takes about an hour to complete.

6

HOW THE ENVIRONMENT AFFECTS THE URINARY SYSTEM

The environment can cause serious urinary-system conditions. Bacteria, viruses, diet, lifestyle choices such as smoking, and chemicals can all affect the kidneys, bladder, and other parts of the urinary system.

BACTERIA AND OTHER ORGANISMS

Bacteria, viruses, and other organisms can cause disease and urinary-system disorders. Urinary-system conditions caused by these organisms include the following:

POST-STREPTOCOCCAL GLOMERULONEPHRITIS

Post-streptococcal glomerulonephritis (pohst-strep-toh-KAHK-ul gloh-mer-yoo-loh-nef-RYE-tiss), or PSGN, occurs after a person has recovered from a bacterial respiratory condition caused by the *streptococcus* (strep-toh-KAHK-uss) bacteria. The bacteria do not directly attack the kidneys. Instead, antibodies produced by the immune system to fight the bacteria build up in the glomeruli and cause the damage. Thanks to antibiotics used to treat strep infections, this kidney condition is now rare.

PSGN usually affects boys between the ages of three and seven. Symptoms of the condition include swelling of the feet and hands, blood in the urine, and reduced urine output. The condition usually goes away after a brief time, and the kidneys usually recover without treatment. In severe cases, permanent and serious kidney damage requires dialysis or a kidney transplant.

URINARY TRACT INFECTIONS

Urinary tract infections (UTIs) usually result from intestinal bacteria that get into the urinary tract. These infections can affect any part of the urinary system. The most common UTIs occur in the bladder and urethra. A UTI In the bladder is cystitis. A UTI in the urethra is urethritis (yoor-ee-THRYE-tiss). A UTI in the kidneys is called pyelonephritis (pye-uh-loh-nef-RYE-tiss). Pyelonephritis can cause serious kidney damage.

UTIs are common medical problems in the United States. Each year, millions of Americans are affected by these conditions. Children and women are much more likely to get UTIs than are men. The elderly, diabetics, and people who are catheterized, especially if the catheter must remain in place for a long time, are also at risk from UTIs.

According to the NIDDK, 3 percent of all girls and 1 percent of all boys suffer a kidney infection during childhood. Some children have problems with their urinary systems that put them at greater risk for these infections. In addition, young children may not have learned how to wipe themselves properly after passing stools or urine, thus creating additional risk of infection.

Women are more likely than men to suffer UTIs because of the structure of their anatomy. In female anatomy, the bladder is lower and the urethra shorter than in males. Harmful bacteria have a shorter distance to travel. Each year, about 11 percent of all United States women get a UTI.

Symptoms of a UTI include pain and burning during urination, the urge to urinate frequently, fatigue, fever, chills, and vomiting. In addition, the urine may be cloudy or bloody or have a strong or foul smell. Most UTIs are easily cured with antibiotics. In severe cases in which the UTI has spread to the kidneys, a patient may need to be hospitalized until the infection is under control. If a patient continues to develop UTIs, it may be an indication of more serious problems, including an abnormality in the urinary tract.

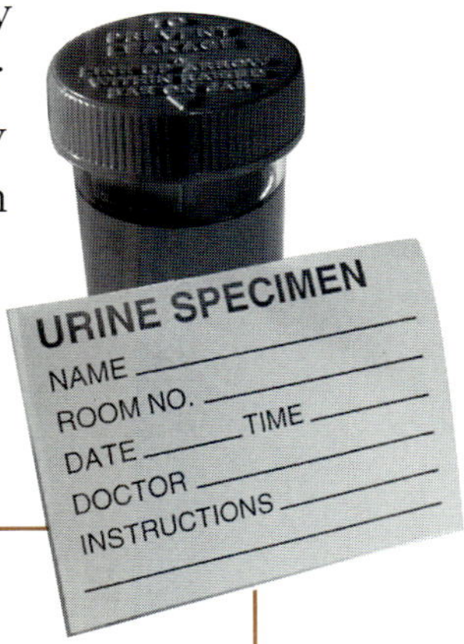

Doctors and medical labs use small, labeled bottles to collect urinary samples for analysis.

Avoiding UTIs

- Drink lots of water. Doctors recommend at least 8 to 10 cups (1.9 to 2.4 liters) each day.
- Women should remember to wipe front to back to ensure that all intestinal bacteria from stool stay where they belong.
- Always urinate when you feel the need. Holding off on urination can cause bacteria to thrive and multiply.
- Take showers instead of baths.

CHEMICALS, SMOKING, DRUGS, AND DISEASE

There are some things that, when taken into the body, can really harm the urinary system. Researchers believe, for example, that some chemicals contribute to cancers of the urinary system. These chemicals include aromatic amines, which are substances used in making dyes, textiles, rubber, and pesticides.

Cigarette smoking greatly increases the risk of certain urinary tract diseases, including bladder and liver cancers. In fact, smokers are five times more likely to develop bladder cancer than nonsmokers are. Certain medicines used to treat other health problems can harm the urinary system. They include certain drugs used in chemotherapy, as well as some over-the-counter medicines.

These two women are increasing their risk of urinary infections and other diseases by smoking.

BLADDER CANCER

Bladder cancer is any cancer that begins in the bladder. This type of cancer usually starts in the cells lining the inside of the bladder. From there, the tumor may grow, invade the bladder muscle, and eventually spread to other nearby organs, such as the uterus, vagina, or prostate. If the cancer reaches nearby lymph nodes, it can spread throughout the body to other organs.

White men over the age of forty are most at risk for bladder cancer. If someone in the family has had bladder cancer, the risk of getting the disease increases. This risk also increases with age: The older a person gets, the more likely he or she is to get the disease. Chronic bladder problems also raise a person's chances of getting this type of cancer.

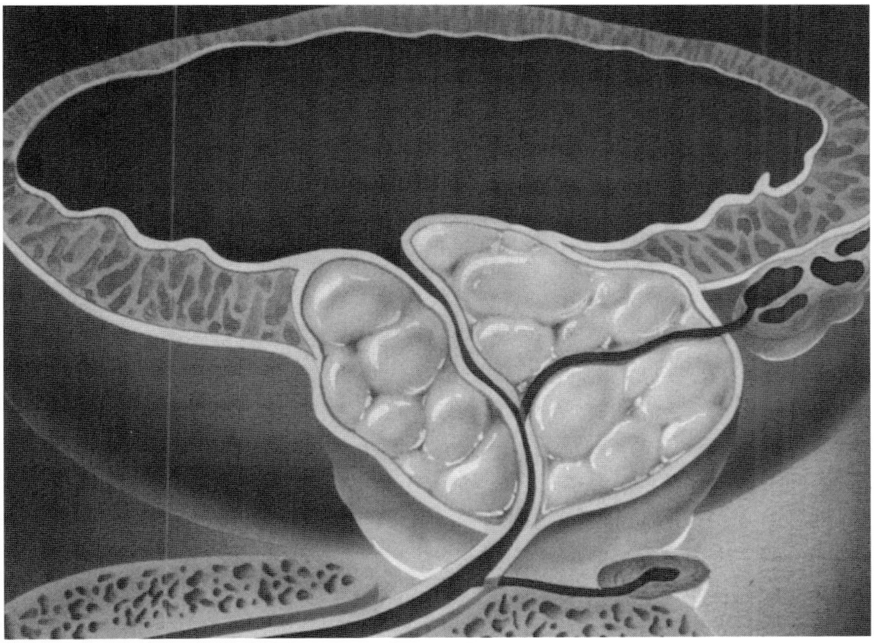

This illustration shows a cancerous growth on the bladder.

Although researchers do not know exactly what causes bladder cancer, they do know that smoking greatly increases the risk. Half of all cases of bladder cancer in males can be attributed to smoking. Chemicals present in cigarette smoke make their way into the bladder where they damage cells.

Another environmental factor that places people at risk for bladder cancer is workplace carcinogens. Workplace chemicals have been linked to about one out of every four cases of bladder cancer. It has been shown that people who work with dye, aluminum, leather, and pesticides are at higher risk for the condition.

Symptoms of bladder cancer include blood in the urine, pain while urinating, and frequent urination or feelings of needing to urinate. Because the disease can be deadly, it is important for people with these symptoms to seek medical attention immediately. As with other cancers, successful treatment depends on whether the tumor is detected at an early or late stage.

TREATING BLADDER CANCER

In order to discuss the severity of a person's bladder cancer doctors identify five different stages, from Stage 0 to Stage IV. Treatment for bladder cancer depends upon where the tumor first formed, what the tumor looks like, and whether it has spread. Stage 0 is a cancerous tumor that is limited to the bladder. This type of bladder cancer has the greatest chance of being successfully treated. At the most serious stage, Stage IV, the cancer has spread to the lymph nodes and other parts of the body.

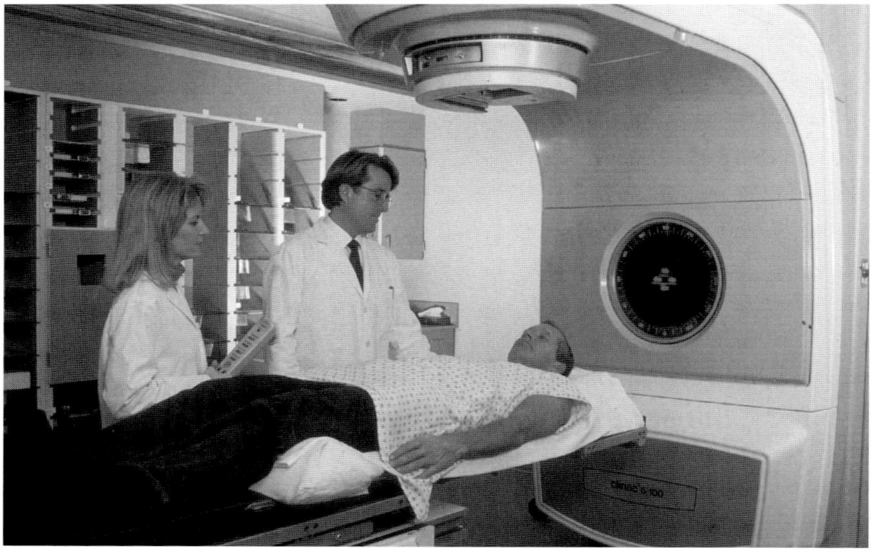

People with bladder cancer must often undergo radiation therapy.

Treatments include surgery, radiation therapy, chemotherapy, and biological therapy. Surgery to remove the cancerous tumor is one of the most common treatments for Stage 0 and Stage I bladder cancer. In some cases, the tumor can be removed by itself. In more serious cases, part or all of the bladder must be removed. When the entire bladder is removed, doctors replace it with a bag, worn inside or outside the body, to hold urine. If the cancer has spread, nearby lymph nodes and other organs may also be surgically removed.

With radiation therapy, doctors use powerful rays of energy to kill the tumor. In some cases, radioactive materials are implanted into the bladder in hopes that they will destroy the cancerous tissues. Eventually, these radioactive substances are naturally flushed from the body.

In some cases, chemotherapy drugs are placed into the bladder through the urethra. In other cases, patients receive an injection of the drugs through a vein. With this method, the drugs travel throughout the body before reaching the bladder.

Biological therapy, or immunotherapy (im-myoon-oh-THERR-uh-pee), is the introduction of a solution of weakened bacteria into the body in hopes of encouraging the immune system to fight the bladder cancer. The bacterial solution is placed into the bladder through the urethra.

The effectiveness of the treatments usually depends on the stage of the bladder cancer when it is first discovered. While Stage 0 and Stage I patients have a good chance of recovery, the recovery rate for patients in Stage III is less than fifty percent. Those with Stage IV cancer are rarely cured.

KIDNEY CANCER

Kidney cancer is any cancer that begins in the kidneys. As cancerous cells multiply, they can spread to nearby organs, including the colon, liver, and pancreas. Sometimes cancer cells break away from the tumor and infect nearby lymph nodes. From there, the cancerous cells can travel to other organs throughout the body, usually the bones and lungs. Each year, about 12,000 Americans die of kidney cancer and more than 28,000 learn that they have the disease.

Doctors aren't sure what causes kidney cancer. They do know, however, that men between the ages of fifty and seventy are more likely to develop this condition than are others. There are also certain risk factors that increase the chances of getting this disease. These risk factors include smoking, obesity, high blood pressure, and exposure to such environmental toxins as coal and asbestos. Dialysis and radiation treatment also put some people at greater risk for kidney cancer.

There are several different types of cancers that affect the kidneys. The most common type of kidney cancer is renal cell carcinoma. It accounts for nine out of ten cases of kidney cancer. This type of cancer begins in the cells that line the small tubules within the kidneys.

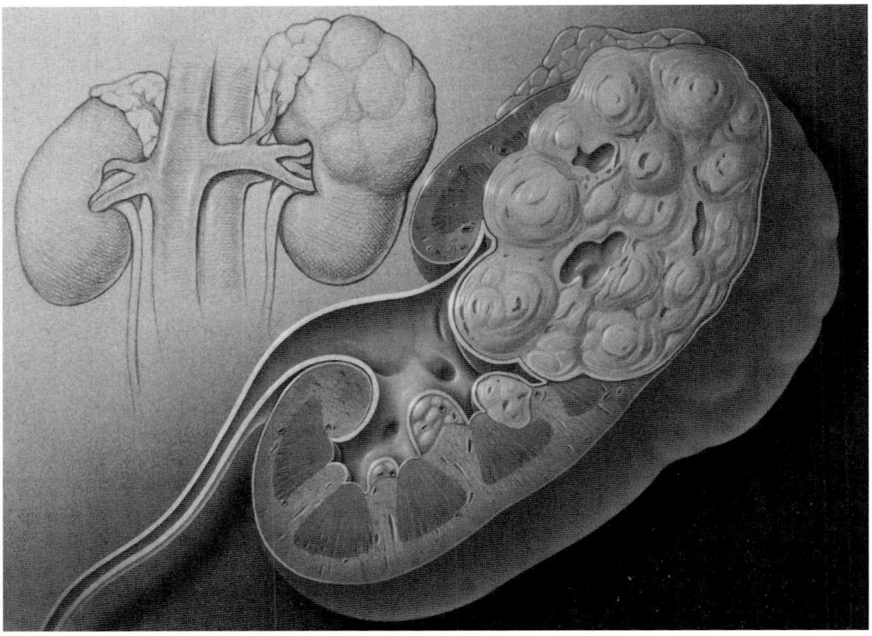

This illustration shows a cancerous growth on the kidney.

Another type of kidney cancer is transitional cell cancer. Transitional cell cancer accounts for between 5 and 10 percent of all kidney cancer cases. This type of cancer begins in the part of the kidneys where urine is collected. Because transitional cell cancer is very similar to bladder cancer, it is often treated in the same way.

Wilm's tumor is the most common type of kidney cancer in children. Although doctors are not sure what causes Wilm's tumor, they think that it may be a congenital disorder. Doctors treat this condition by removing the cancerous kidney.

Patients with early kidney cancer may have few symptoms. Later, symptoms of the disease include blood in the urine and a lump in the kidney area. Other less common symptoms are fatigue, weight loss, frequent fevers, and constant pain in the back or side. People with any of these symptoms should see a doctor immediately. The earlier the cancer is detected, the better the chance of recovery.

Doctors describe the growth and spread of kidney cancer by talking about what stage the cancer is in. Kidney cancer has four stages, Stage I to Stage IV. A Stage I kidney cancer is a small tumor that has not spread beyond the kidneys. A Stage IV kidney cancer is one that has spread to the fatty tissue that surrounds the kidneys, as well as the lymph nodes or other organs.

The most common treatment for early-stage kidney cancer is surgery. During kidney surgery, called a *nephrectomy* (nef-REK-tuh-mee), a surgeon removes part or all of a kidney. In addition, the adrenal gland, surrounding tissue, and nearby lymph nodes may be removed at the same time. Treatment for transitional cell carcinoma may require the removal of the ureters and part of the bladder, as well.

Other kidney cancer treatments include radiation therapy, chemotherapy, biological therapy, and arterial embolization. With arterial embolization, doctors inject pieces of spongelike material into the patient's kidney. The sponge clogs the main renal vein, cutting off the tumor's supply of blood and nutrients. As a result, the tumor shrinks.

DRUGS AND MEDICATIONS

Consistent, long-term use of some over-the-counter medicines can cause kidney damage. One condition caused by painkillers is analgesic nephropathy. An *analgesic* (an-al-JEEZ-ik) is a medicine used to stop pain. *Analgesic nephropathy* is a chronic deterioration of kidney function that can eventually lead to kidney failure. Medicines that can cause harm when taken excessively over a number of years include aspirin, acetaminophen (uh-see-toh-MIN-uh-fen), and ibuprofen. Painkillers that combine these types of medicine with codeine or caffeine are particularly dangerous to the kidneys.

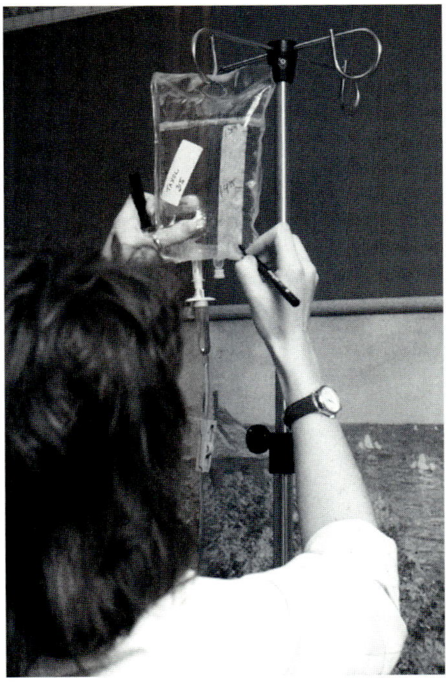

A hospital worker labels a vial used to dispense chemotherapy drugs to a patient intravenously.

One drug that is linked to kidney cancer is phenacetin (fee-NASS-uh-tin). More than twenty years ago, this drug could be found in over-the-counter pain relievers. Because of the risk that it poses to people's health, however, phenacetin is no longer available in the United States. Another type of medicine that has been linked to kidney cancer is diuretics, used to treat high blood pressure and heart failure.

Unfortunately, treatment for some types of cancer can lead to an increased risk of urinary cancers. Radiation for cervical cancer, for example, raises the risk of bladder cancer. Cyclophosphamide (sye-kloh-FOSS-fuh-myde), a chemotherapy drug used to treat lymphoma, leukemia (loo-KEE-mee-uh), breast cancer, and other types of cancer, also increases the chances of getting this serious disease.

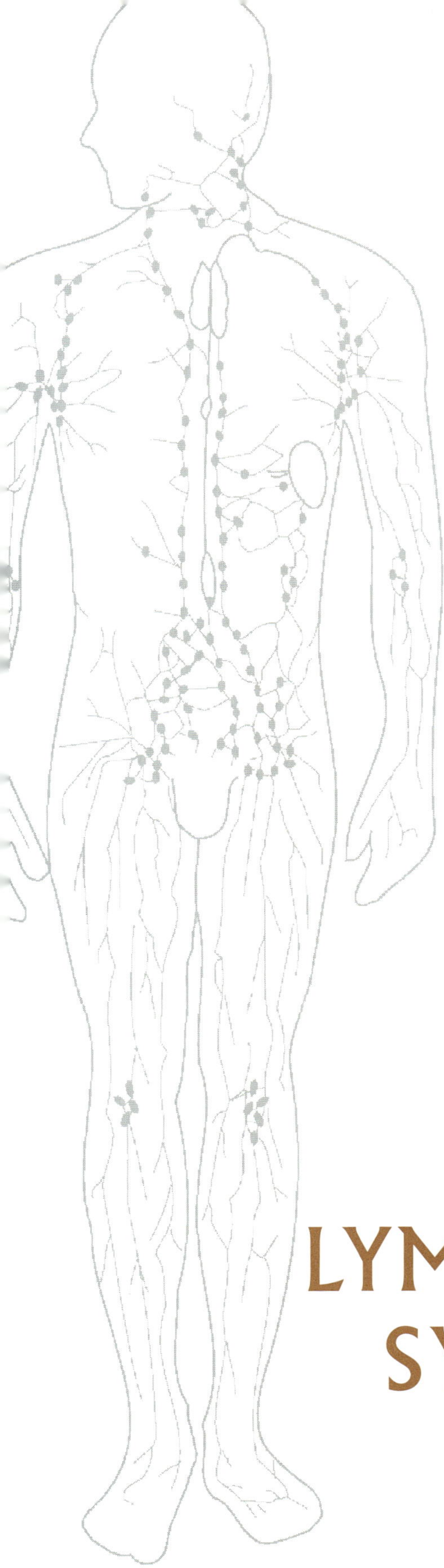

7

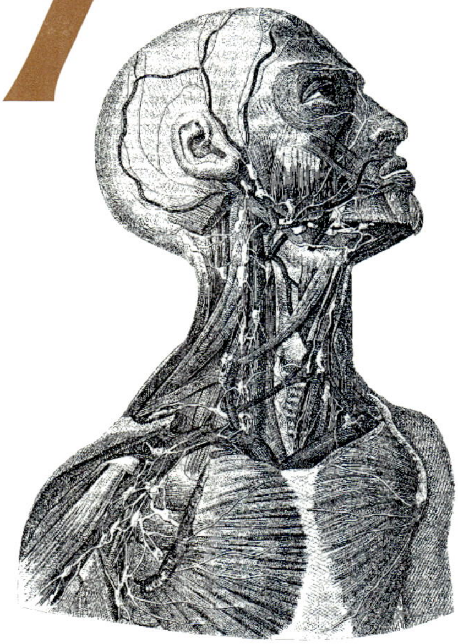

THE LYMPHATIC SYSTEM

The job of the *lymphatic* (lim-FAT-ik) *system* is to filter excess water, proteins, and fat from the body's cells and tissues and return them to the blood vessels. Without the lymphatic system, the body would become swollen and puffy with these unneeded fluids. The lymphatic system plays important roles in other body systems, as well. For example, the lymphatic system works with the immune system to fight infection and cancer. The lymphatic system is also closely related to the circulatory system because *lymph,* the fluid that transports the water and proteins, is produced in the blood and returns to blood.

PARTS OF THE LYMPHATIC SYSTEM

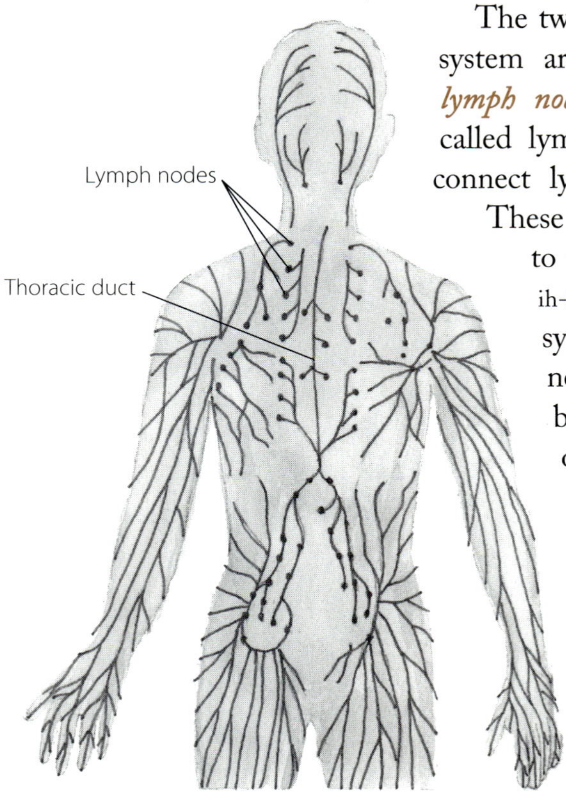

The two chief parts of the lymphatic system are the *lymph vessels* and the ***lymph nodes.*** The lymph vessels, also called lymphatics, are tiny vessels that connect lymph nodes to one another. These vessels are similar in structure to the veins and capillaries (KAP-ih-leh-reez) of the circulatory system. Lymph vessels, found in nearly every part of the human body, transport lymph throughout the body. The smallest of the lymph vessels are the lymph capillaries. Lymph capillaries connect to larger lymph veins, which in turn connect to one of two larger vessels called lymph ducts. The largest lymph vessel in the body is the *thoracic* (thor-RASS-ik) duct. The thoracic duct receives lymph from most of the left side of the body. The right lymphatic duct receives lymph from the right side of the body. Both ducts empty lymph into large veins in the upper chest.

Lymph is the clear, watery fluid that comes from blood and carries excess body liquids. It is made up of water, proteins, salts, and other substances but contains no blood cells. Lymph is moved through the body by muscle contractions. Tiny valves on the lymphatics keep the lymph moving in one direction—away from cells and tissues and toward the blood vessels. After passing through the lymphatics, lymph returns to the blood.

> **Fast Fact**
>
> Lymph nodes are tiny. The largest are only 1 inch (2.5 centimeters) in size.

Lymph nodes, also called lymph glands, are small, bean-shaped organs. They can be found everywhere in the body. In some parts of the body, the nodes can be easily detected. When people have an infection or illness, they often notice swollen lymph nodes in the neck, under the armpits, in the groin, and on the back of the head above the hairline.

Lymph nodes produce some types of *lymphocytes* (LIM-foh-sites), or white blood cells. These lymphocytes manufacture antibodies to fight infections and disease. As lymph passes through the nodes, the tiny organs filter out harmful substances that can make people ill. Then the lymphocytes attack and destroy these organisms.

Other organs of the body are also part of the lymphatic system. An important lymphatic organ is the spleen. As part of the immune system, the spleen contains antibodies that fight infection. The spleen also contains *macrophages* (MAK-roh-fay-jez), white blood cells that filter damaged cells from the bloodstream, destroy them, and send them on their way to be cleared out of the body.

Other parts of the body that play key roles in the lymphatic system include the following:

- *Bone marrow:* Lymphocytes are produced here. B cell lymphocytes mature here, but T cell lymphocytes mature in the thymus (THYE-muss) gland.
- *Thymus gland:* This is a small organ found near the base of the neck. The thymus gland of a child is made up mostly of lymphatic tissue. After puberty, this tissue is replaced by fat. T cell lymphocytes mature here.
- *Tonsils:* These are two clumps of lymphatic tissue found on the sides of the throat. Tonsils trap bacteria and viruses that enter the body through the mouth.
- *Adenoids:* These are clumps of lymphatic tissue located behind the nose that trap bacteria and viruses that enter the body through the nasal passages. The adenoids also produce antibodies to fight infection.

Health and History: Thomas Bartholin

The first person to discover the presence of the lymphatic system in the human body was Danish scientist Thomas Bartholin (1616–1680). In 1652, Bartholin used the cadavers (dead bodies) of two criminals to discover the location of the thoracic duct in humans. He was the first person to recognize the lymphatic system as an entirely separate body system.

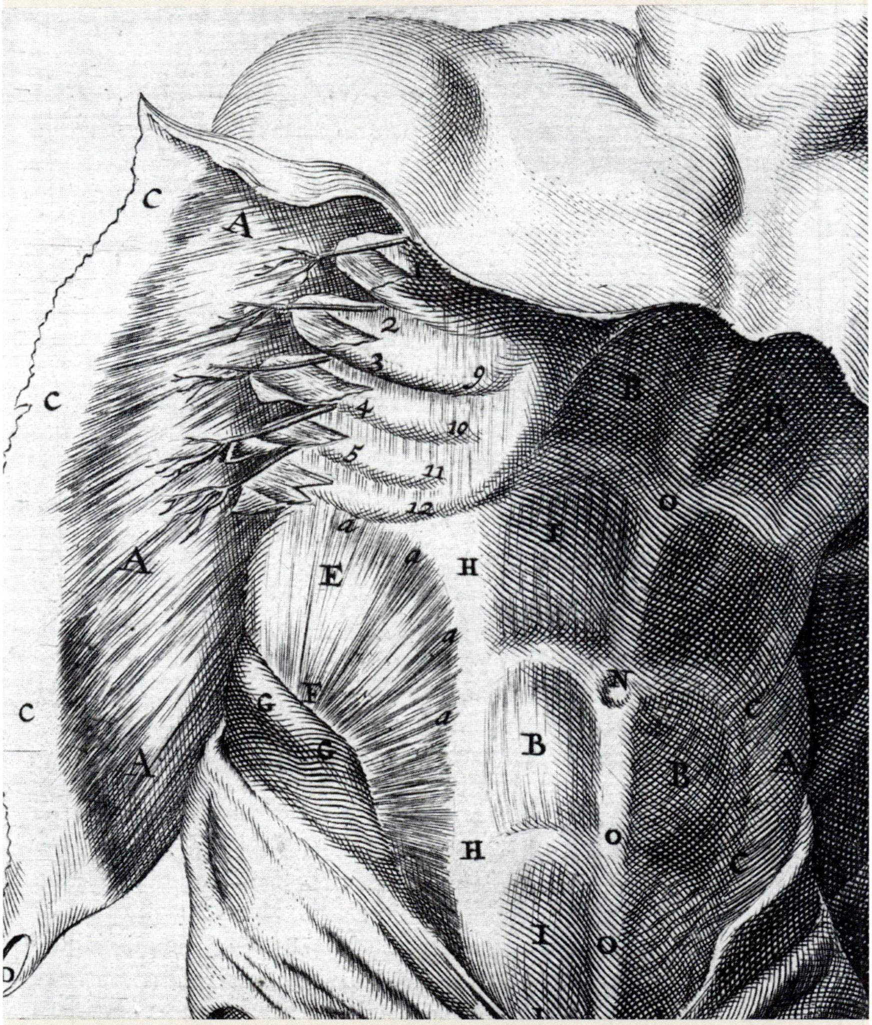

This illustration depicting a dissected male torso is from *Anatomia Reformata* by Thomas Bartholin, written and published in the 1600s.

8

LYMPHATIC SYSTEM DISORDERS

The lymphatic system plays an important role in filtering out foreign substances that can damage the body. Swelling lymph nodes are often the first indicator that a person is unwell. A number of conditions can affect the lymphatic system, including lymphedema and ruptured spleen.

LYMPHEDEMA

Lymphedema (lim-feh-DEE-muh), also known as lymphatic obstruction, is a condition in which the lymphatic vessels swell and become blocked, causing lymph to build up in the fatty tissue just below the skin. This buildup causes swelling, especially in the arms and legs. Lymphedema can result in infection of the lymph nodes, which in turn can impair healing of wounds and prevent proper circulation.

There are two types of lymphedema, primary and secondary. Primary lymphedema may be present at birth or may occur at puberty or in adulthood. It may be the result of damaged lymph vessels, missing lymph nodes, or circulatory conditions. However, the cause of many primary lymphedema cases is unknown.

The causes of secondary lymphedema are better understood. Most cases are blockages in or damage to the lymphatic system as a result of surgery, radiation, trauma, or infection. Certain types of surgery are known to raise the risk of getting lymphedema. Breast cancer surgery is the top lymphedema-causer. Other operations that increase the chances of getting lymphedema include gynecological (gye-neh-koh-LAH-jik-ul), head or neck, prostate, and bladder or colon surgeries. Lymphedema may develop immediately after the surgery or many months—even years—later. Radiation therapy and some medications are also known to increase the risk of getting lymphedema.

In addition to swelling of the arms and legs, other symptoms of lymphedema include tightness of the skin, loss of flexibility in the wrists and ankles, and difficulty wearing clothing and jewelry. If the condition goes untreated, proteins can build up in the tissues and harden. At its most severe, the swelling caused by lymphedema becomes irreversible, and those suffering from the condition may have to undergo reconstructive surgery to reverse the effects. Lymphedema may even lead to a rare form of lymphatic cancer called lymphangiosarcoma (lim-fan-jee-oh-sar-KOH-muh).

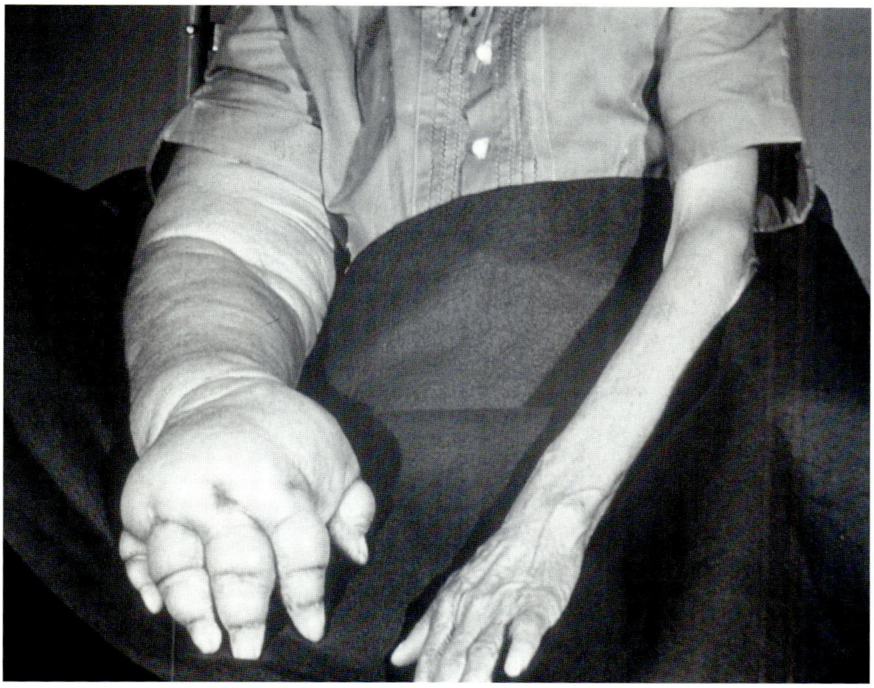

This person's left arm has undergone surgery to reverse the effects of lymphedema. The right arm still shows the extreme buildup of fatty tissue and swelling characteristic of the disease.

While lymphedema can be a temporary condition, many cases are chronic. This means that patients must deal with the disease for the rest of their lives. Before doctors can treat lymphedema, they must first determine what caused the condition. If it has been caused by bacteria, for example, doctors can treat it with antibiotics. Other treatments include using stockings and pumps that compress the swollen areas; elevating swollen limbs; and tending to the affected skin areas. In some cases, surgery is necessary, but operations for this condition are usually experimental and not generally successful.

RUPTURED SPLEEN

Often the result of a traumatic injury or an accident, a *ruptured spleen* is a serious problem. When the spleen is damaged, large amounts of blood can spill into the abdomen, causing pain and very low blood pressure. If the abdominal bleeding is not quickly stopped, the injury can result in death.

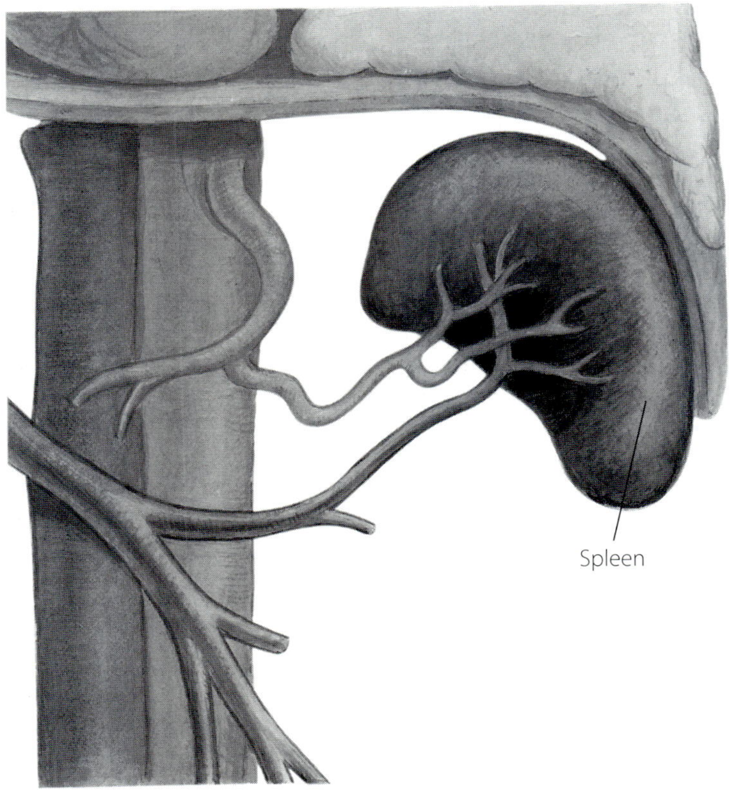

This illustration shows the location of the spleen in the body.

In the worst cases, surgery must be performed immediately to repair the damage and prevent shock and death. In most cases, doctors remove the spleen. This procedure is called a *splenectomy* (spleh-NEK-tuh-mee). A splenectomy increases a person's risk of suffering infections. Those who have had their spleen removed must take extra care to stay up to date on vaccinations to prevent infection, and they may have to take antibiotics to stave off bacteria.

9

HOW THE ENVIRONMENT AFFECTS THE LYMPHATIC SYSTEM

Environmental agents that enter the body can seriously damage the lymphatic system. These agents include bacteria and other harmful organisms, as well as cigarette smoke and hazardous chemicals.

BACTERIA, VIRUSES, AND OTHER HARMFUL ORGANISMS

One of the chief jobs of the lymphatic system is to kill off any foreign organisms that can harm the body. These disease-causing agents can also damage the lymph system:

LYMPHADENOPATHY

Lymphadenopathy (lim-fad-eh-NAH-puth-ee) is a general term for any swelling of the lymph nodes. The condition, especially in children, is usually the result of an infection in the surrounding area. In some cases, the swelling may be caused by a serious medical condition, such as cancer. When lymphadenopathy occurs in glands around the body, it indicates a systemic problem.

LYMPHADENITIS

Lymphadenitis (lim-fad-eh-NYE-tiss) is a swelling of a lymph node that is usually caused by a bacterial, fungal, or viral infection. The bacteria that most often cause this condition are *staphylococcus* (staff-ih-loh-KAHK-uss) and *streptococcus.* The swelling occurs as the node produces more and more lymphocytes to fight the infection. Lymphadenitis may also be caused by cancer cells present in the particular node.

Symptoms of lymphadenitis include swelling, hardness, tenderness, and warmth or redness of the skin above the infected nodes. This condition most often occurs in the nodes of the neck, especially when the patient is a child. This could indicate an infection of the mouth, throat, or scalp. Doctors may take a biopsy of the swollen lymph nodes to determine what is causing the infection.

Doctors treat the condition with antibiotics, as well as pain-relief and anti-inflammatory drugs to ease aches and swelling. Although patients usually recover from lymphadenitis quickly with the proper treatment, complications can result. If the infection spreads into the bloodstream, the condition can even prove fatal. It is important for people with swollen glands to seek immediate medical attention.

LYMPHANGITIS

Lymphangitis (lim-fan-JYE-tiss) is a swelling of the lymph vessels. Like lymphadenitis, this condition is often the result of a strep or staph infection of the skin. In many cases, lymphangitis is a sign that an infection is spreading.

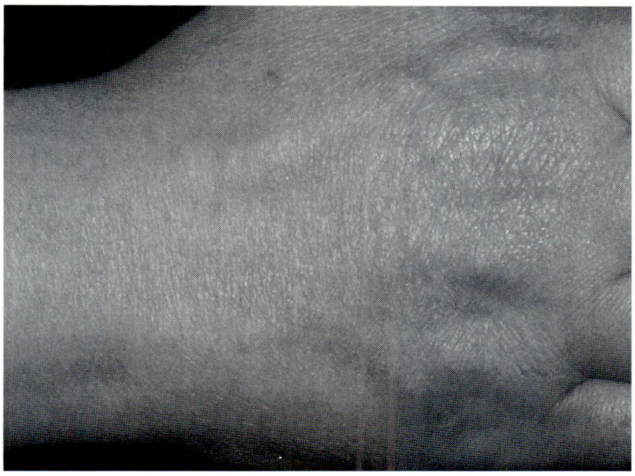

This photo shows swelling in the hands caused by lymphangitis. This disease can cause swelling in other parts of the body also.

Symptoms of the condition include pain, red streaks that stretch from the infected area to the armpits or groin, high fever, headaches, and loss of appetite. Like lymphadenitis, lymphangitis can spread rapidly. If the infection spreads into the bloodstream, it can cause serious illness. Doctors use antibiotics to cure the condition.

LYMPHATIC FILARIASIS

Lymphatic filariasis (fil-uh-RYE-uh-siss) is a condition that develops when worm larvae are injected into the lymphatic system through the bites of mosquitoes. It usually takes many mosquito bites before lymphatic filariasis develops. Although the condition does not occur in the United States, it affects more than 120 million people a year in other parts of the world. The condition is most common in Southeast Asia, India, and Africa.

Once the larvae have been injected into the lymphatic system, they reproduce and spread throughout the body. The disease progresses as the larvae mature into worms in the lymph system, causing swelling. Adult worms can live inside the body for seven years.

At first, people infected with filarial larvae exhibit no symptoms. Eventually, however, the larvae can cause serious damage to the lymph system and kidneys. Later symptoms include edema, an increase in other infections as the lymph system weakens, and hard, thick skin. This thickening of the skin is known as *elephantiasis* (el-eh-fan-TYE-uh-siss). It can leave patients scarred, disfigured, and disabled.

While filarial larvae can be killed with medication, there is no medicine to kill adult filarial worms. People with adult worms must take extra precautions to avoid spreading the infection to others. They can also take certain steps to lessen the swelling associated with the condition. Once a year, infected patients should receive a dose of medicine to kill any new worm larvae that have developed.

CAT SCRATCH DISEASE

Cat scratch disease (CSD), also known as cat scratch fever, is an infection caused by a bacterium called *Bartonella henselae* (bar-tuh-NELL-uh HEN-suh-lye). This bacterium, usually transmitted through the bite or scratch of an infected cat, causes lymph nodes to swell near the site of the wound. The bacteria can also be spread if the infected animal's saliva enters the eye or cuts in the skin.

Symptoms of CSD usually begin two to three weeks after contact with the infected animal. These include swollen lymph nodes, fever, headaches, and fatigue. Sometimes, the swollen nodes will break through the skin and drain. In severe cases, the spleen may become enlarged. Doctors treat this serious infection with antibiotics to kill the bacteria. To avoid CSD, people, especially children, should avoid cat bites and scratches and should always wash their hands after playing with cats.

SPLENOMEGALY

Splenomegaly (spleen-uh-MEG-uh-lee), also known as an enlarged spleen, is a condition that is most often caused by viral infection. One virus associated with splenomegaly is the Epstein-Barr virus—the same one that causes the infectious disease mononucleosis (mahn-oh-noo-klee-OH-siss). Other viruses that cause the condition include those that cause hepatitis, tuberculosis (too-ber-kyoo-LOH-siss), and malaria. Splenomegaly may be caused by certain other medical conditions, including leukemia, lymphoma, and cirrhosis.

As the spleen becomes enlarged, it absorbs more blood cells. This can negatively affect the body by causing anemia, increased risk of infections, and bleeding problems. As the spleen absorbs more cells, it continues to grow even larger. Eventually, it begins to absorb normal cells, as well as abnormal cells.

There are few symptoms for this condition. If the spleen becomes large enough to press against the stomach, a person may experience a full sensation, even after eating small amounts or no food at all. Another symptom is back and abdominal pain, followed by pain in the left shoulder. Doctors often detect splenomegaly during a physical and can confirm their diagnosis by performing an X-ray, MRI, or other imaging test to see inside the body.

Splenomegaly can be especially dangerous if the spleen is ruptured. Because the enlarged spleen contains abnormal amounts of blood, a person with a damaged spleen can suffer a huge loss of blood. People who know that they are suffering from splenomegaly should avoid physical activity until their condition has been treated.

CUTANEOUS ANTHRAX

People become infected with *cutaneous* (kyoo-TAY-nee-uss) *anthrax* when spores, or cells, from anthrax bacteria enter the body through wounds or breaks in the skin. The bacteria are transported throughout the body by the lymph system and can eventually enter the bloodstream and cause serious damage.

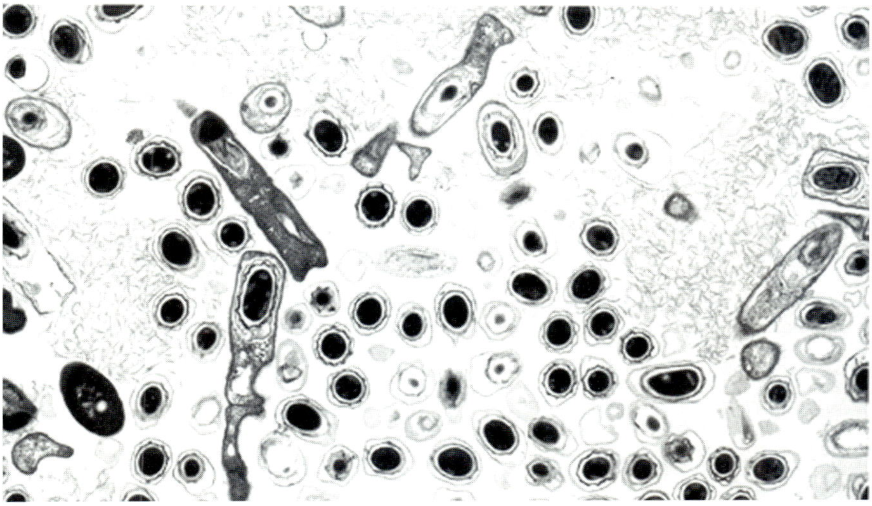

This photograph, taken with a microscope, shows anthrax spores in the bloodstream.

Cutaneous anthrax is the most common type of anthrax. About 95 percent of all people infected with the bacteria have this type of anthrax. According to the CDC, the disease is usually contracted through occupational exposure to infected livestock or livestock products such as hair, hides, bones, or wool. It is more common in agricultural regions of the world.

The first symptom of cutaneous anthrax infection is an itching at the point of infection. The itching is followed by a blister, which develops into a painless sore covered with a black scab. Lymph nodes near the site of the infection may become swollen. It is important to get this type of anthrax treated. In two out of ten cases, the infection can spread to the blood and cause illness and death. Doctors can effectively treat cutaneous anthrax with antibiotics that kill the bacteria.

Anthrax has been in the news lately. Since 2001, Americans have become concerned about the potential use of the bacterium as a weapon of terror. The use of bacteria and other biological agents in this manner is known as *bioterrorism.* Experts say, however, that anthrax could probably not be used as a weapon on a large scale because there is no effective delivery system.

TONSILLITIS

Tonsillitis (tahn-sil-LYE-tiss) is the swelling of the tonsils, the masses of lymph tissue found in the throat. The condition, which most often affects small children, usually occurs as a result of bacterial or viral infection. The most common type of tonsillitis is caused by *streptococcus* bacteria.

A doctor can test a patient for tonsillitis by performing a throat-swab culture to check for bacteria or viruses that might be present. The doctor rubs the back of the patient's throat with a cotton swab. The swab is then tested for harmful organisms.

Symptoms of this condition include a sore throat, fever, difficulty swallowing, and swollen lymph glands in the neck. The infection can spread and cause *pharyngitis* (faar-in-JYE-tiss), an inflammation of the pharynx. Another serious complication of tonsillitis is called *peritonsillar* (perr-ih-TAHN-sil-ler) *abscess.* This occurs when pus and dead blood cells collect and block the airways. This infection can spread into the neck and chest and eventually prove fatal. Doctors usually treat an abscess by draining it before it can cause serious problems.

Doctors usually treat tonsillitis with antibiotics. While recovering from this illness, a patient should get plenty of rest and drink lots of fluids. Pain relievers to lower fever and reduce discomfort can also be used. After two or three days, the patient should begin to feel better.

If a child continues to get tonsillitis repeatedly, surgery to remove the tonsils may be necessary. This procedure is called a *tonsillectomy* (tahn-sil-LEK-tuh-mee). Most patients can leave the hospital the same day that this procedure is done. In some cases, the doctor may also remove the adenoids at the same time.

CANCER AND THE LYMPHATIC SYSTEM

Like other body systems, the lymphatic system can be affected by cancer. A number of cancers originate in the body's lymphatic system. Other cancers spread from their point of origin throughout the body via the lymphatic system.

Lymphoma (lim-FOH-muh) is the term for any type of cancer that begins in the lymphatic system. Lymphoma occurs when abnormal lymph cells begin to multiply rapidly, creating masses of unneeded tissue that form tumors. From the lymph nodes, the cancer may then spread to other parts of the body.

About 5 percent of all cancer cases in the United States are lymphoma. According to the ACS, 60,900 Americans learn each year that they have lymphoma. About 1,700 of these patients are children. Although the condition is serious, the chances for successful treatment are high. Most children, for example, are cured by treatment.

> **Fast Fact**
>
> Parents exposed to ionizing radiation in the workplace may put their children at greater risk for childhood lymphoma. Those who work around such radiation include X-ray technicians, miners, and nuclear and aircraft workers.

Medical researchers believe that a number of factors cause lymphoma. Viral and bacterial organisms have been linked to this type of cancer. Certain chemicals may also cause lymphoma. It is known for certain that some people are more at risk for lymphatic cancer than are others. White males with immune system problems, for example, are at a higher risk. Immune system problems could be caused by genetic conditions, HIV, or drugs that suppress the immune system after organ transplants. In addition, children who have received radiation treatment or chemotherapy for other types of cancer may also be at higher risk.

Lymphomas are divided into two main categories: Hodgkin's lymphoma and non–Hodgkin's lymphoma (NHL). Hodgkin's lymphoma is much less common than NHL. It represents just 11 percent of all lymphomas. The cancerous cells of Hodgkin's and NHL look very different from one another. To determine which type of lymphoma a person is suffering from, doctors perform a biopsy of an affected lymph node. To see how far the cancer has spread, they can do blood tests, CT tests, spinal taps, bone marrow biopsies, ultrasounds, and MRIs. After performing a battery of tests, doctors can decide how best to treat the condition.

> **Fast Fact**
>
> Hodgkin's disease is named for British physician Thomas Hodgkin (1798–1866). In 1832, he became the first doctor to describe the condition.

HODGKIN'S LYMPHOMA

Hodgkin's lymphoma, also known as Hodgkin's disease, is a cancer of the lymph nodes, spleen, liver, and bone marrow. The cause of Hodgkin's disease is not known. This disease most commonly affects those in early or late adulthood (ages fifteen to thirty-five and over fifty-five). The incidence of this cancer in the United States has decreased over the past few decades. Today, about two out of every 100,000 Americans suffers from Hodgkin's disease.

The first symptoms of the disease include painless swelling of the lymph nodes in the neck, armpits, or groin or above the collarbone. Other symptoms include fatigue, weight loss, hives, fever, and night sweats. If the cancer is in the thymus, a patient may also experience coughing and shortness of breath.

Doctors diagnose this disease by identifying a certain type of cancerous cell unique to the condition. These cells are called Reed-Sternberg cells. Hodgkin's is very different from NHL in the way that it spreads throughout the body. Unlike NHL, Hodgkin's spreads in a predictable and specific way. It usually develops in one region and then spreads to other parts of the lymph system before spreading to organs and other body systems.

NON-HODGKIN'S LYMPHOMA

NHL is any cancer that originates in the lymphatic system that is not Hodgkin's disease. This type of lymphoma is more common than Hodgkin's disease. It affects three out of every 10,000 Americans. Each year, about 500 children in the United States are diagnosed with the disease. NHL is the fifth most common cause of cancer deaths in the United States. About 25,700 people die from the condition each year.

There are at least thirty different types of NHL. Each type can be categorized as either high grade, intermediate grade, or low grade, depending upon how quickly the tumor grows and what the patient's chances for recovery are. A high-grade tumor, for example, grows rapidly and requires aggressive treatment. One type of high-grade tumor is Burkitt lymphoma. A low-grade tumor grows more slowly, and the chances for successful treatment are much greater.

Michael Decker climbed Mount McKinley, the highest peak in North America, despite suffering from non-Hodgkin's lymphoma. His effort raised awareness of the disease and funds for cancer research.

The primary symptom of NHL is a swelling of the lymph nodes, particularly in the armpits, groin, or neck. Other symptoms include fever, night sweats, and weight loss. From the lymph nodes, the cancer can spread to any part of the body. Doctors diagnose NHL by looking for the malignant growths of certain types of lymphocytes that collect in the lymph nodes.

Some people are more prone to NHL than others. White males are especially at risk, and that risk increases as they get older. People with certain medical conditions, such as Down's syndrome, rheumatoid (ROO-muh-toyd) arthritis, HIV infection, and lupus, have an increased risk of getting NHL. Viruses, bacteria, and diets high in fats and red meat can also increase the risk for this type of cancer.

Another risk factor for NHL is exposure to certain chemicals. In contrast to Hodgkin's disease, the number of people suffering from NHL each year has increased in recent decades. Some experts believe that this increase is caused by the excess of chemicals that people come into contact with today. Every day, people are exposed to chemicals at work, at home, and elsewhere.

Chemicals that have been linked to cancer of the lymphatic system include arsenic, lead, vinyl chloride, asbestos, and certain pesticides. Studies have shown that when small children come into contact with pesticides, they have an increased risk of getting non-Hodgkin's lymphoma. Other studies have shown that farmers who use certain herbicides also have a higher rate of lymphoma and other health conditions.

One chemical that has been linked to lymphoma is Agent Orange, an herbicide used by the United States military in South Vietnam during the Vietnam War (1964–1975). Agent Orange was used to destroy the forests where enemy soldiers could hide. After the war, Vietnam veterans began having numerous health problems. The incidence of these problems was much higher among veterans than among other United States populations. Health officials noted that veterans more frequently developed Hodgkin's disease, NHL, and other health conditions. In 2002, the United States and Vietnam signed an agreement to work together researching Agent Orange and its effects on the human body and the environment. The two countries will jointly fund research on the toxic substance, and scientists from around the world will share knowledge, data, and results of prior studies.

A United States Air Force plane sprays defoliants along a South Vietnamese highway in 1966. Chemical defoliants such as those used in the Vietnam conflict have been associated with Hodgkin's disease and other cancers.

Another chemical suspected of causing tumors of the lymphatic system is acrylonitrile (uh-kril-oh-NYE-tril). This chemical is used to make dinnerware, containers, toys, telephones, luggage, and other goods from plastic, rubber, and other materials. Research has shown that workers at rubber plants who are exposed to this chemical have a higher incidence of lymphatic tumors than the general population.

DETECTING LYMPHOMA

Doctors use a number of methods to detect and diagnose lymphoma. In a *lymph node biopsy,* all or part of a lymph node is removed and then examined under a microscope for the presence of cancerous cells.

Imaging tests, including X-rays, CT scans, and MRIs, take pictures of the inside of the body. They allow doctors to see the size of a tumor and determine whether or not it has spread.

> **Fast Fact**
>
> Dye injected into the body during a lymphangiogram can remain in the lymph nodes for up to two years.

In a *lymphangiogram* (lim-FAN-jee-oh-gram), dye is injected into the body, usually in the feet or hands. The dye, which is visible through the skin, helps doctors locate normally invisible lymph vessels or nodes. At this point, doctors make a small cut in a large lymph vessel and insert a needle or catheter there. More dye is injected into the lymph system, and X-rays of the body are then performed.

A *laparotomy* (lap-uh-RAHT-uh-mee) is a surgical procedure. Doctors performing this procedure by cutting open the abdomen to examine the lymph nodes.

STAGING AND TREATING LYMPHOMA

In order to discuss how severe a person's lymphoma is, doctors identify four different stages. Here is how Hodgkin's lymphoma is staged:

- **STAGE I:** Hodgkin's lymphoma involves just one lymph node region.
- **STAGE II:** Hodgkin's lymphoma involves two lymph nodes on the same side of the diaphragm, the muscle that separates the chest and the abdomen.
- **STAGE III:** Hodgkin's lymphoma involves lymph nodes on both sides of the diaphragm.
- **STAGE IV:** Hodgkin's lymphoma has spread to tissues and organs outside the lymphatic system.

How doctors treat lymphoma depends upon a number of factors, including the type of lymphoma, which stage it is in, where the cancer originated, and the patient's age and health. The two main treatments for lymphoma are radiation and chemotherapy. Radiation treatment uses high energy X-rays to shrink the cancerous tumors.

Radiation is often used alone when the cancer is confined to one specific region of the body. In cases where the cancer has spread, radiation is used in combination with chemotherapy.

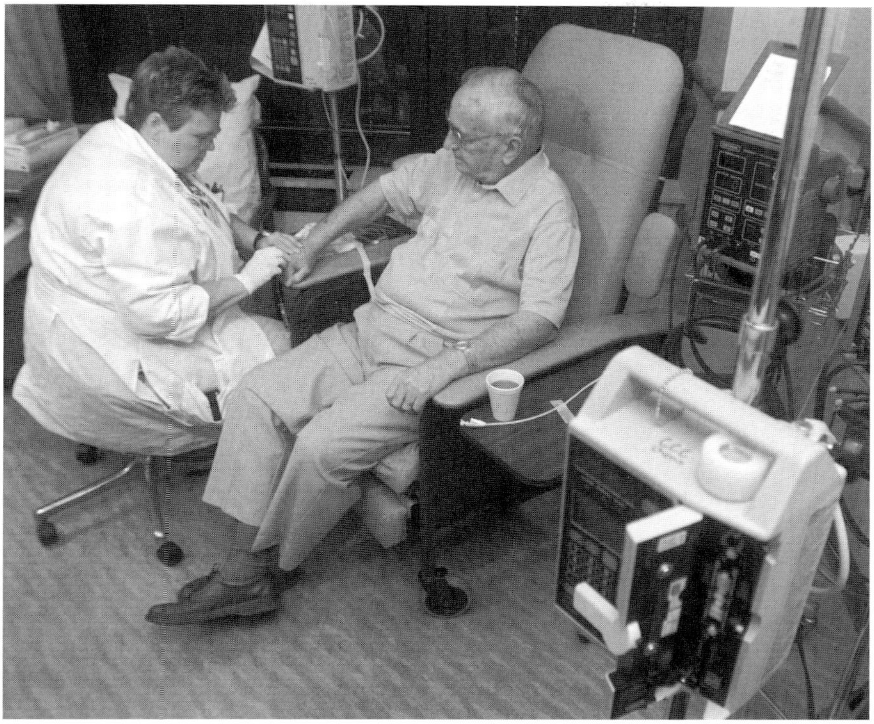

A nurse administers an experimental drug to a patient in an effort to combat lymphoma.

Although these treatments for lymphoma save lives, they sometimes have serious medical side effects. Side effects of radiation treatment include fatigue, skin problems, slowed growth in children, and thyroid (THYE-royd) problems. Radiation treatment has also been linked to an increased risk of breast, lung, stomach, and bone cancer. Patients with Hodgkin's disease who are treated with radiation run an increased risk of getting NHL. Chemotherapy can lead to anemia, bleeding problems, and an increased chance for serious infections. It has also been linked to an increased risk of leukemia.

Some patients do not respond to traditional radiation and chemotherapy. For these patients, doctors may try higher doses of chemotherapy in conjunction with other treatments, including bone marrow transplants or stem cell transplants. The goal of a bone marrow transplant is to replace bone marrow that may have been destroyed during chemotherapy. During this procedure, healthy new marrow from a donor is injected into the bone.

Some doctors choose to perform stem cell transplants after chemotherapy. *Stem cells* are cells in the body that produce healthy new cells and tissue. In this procedure, doctors remove healthy stem cells from a person who is about to begin chemotherapy. After chemotherapy, the stem cells are injected back into the body, spurring new cell growth.

Immunotherapy is another treatment for lymphatic cancer. Surgery may also be an option for some lymphoma patients. The prognosis for lymphoma patients varies, depending upon the type and severity of the cancer.

10

KEEPING THE DIGESTIVE, URINARY, AND LYMPHATIC SYSTEMS HEALTHY

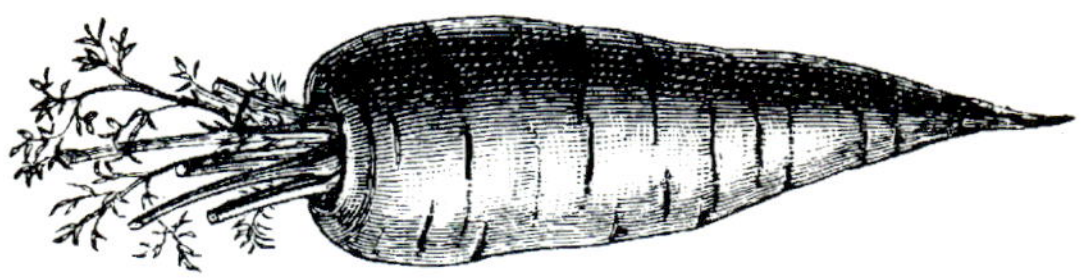

There are many ways that people can keep their bodies as healthy as possible. Eating right, exercising, and making sensible choices about drugs and alcohol are all ways to help the digestive, urinary, and lymphatic systems.

DIET

What people choose to eat can make a big difference in how their bodies work. It is important for people to have healthy, well-balanced diets to ensure that the body gets all the vitamins, minerals, and other nutrients that it needs. The digestive system is greatly affected by what people put into their bodies. A diet that is high in vegetables and fruits can reduce the risks of such digestive-system disorders as cancer of the esophagus.

A father and son select fruits and vegetables at an open-air market. Foods high in fiber are necessary for good intestinal health.

Fiber, a complex carbohydrate found in plants, is great for the digestive system. Not only does it prevent constipation and keep people regular, it also reduces the risk of more serious conditions, including diverticular disease, IBS, and polyps. Fiber is found in vegetables, fruits, whole grain products, and many other foods, but beware: Too much fiber can cause flatulence!

Two Types of Fiber

There are two types of fiber: soluble and insoluble. Soluble fibers dissolve in water, while insoluble fibers do not. Both types of fiber can be beneficial to the human body. Soluble fibers are thought to lower blood pressure, reducing the risk of heart disease and diabetes. Insoluble fibers are good for digestion. They can help prevent colon cancer and diverticulitis.

People suffering from some digestive conditions can care for themselves by carefully watching what they eat. People with cirrhosis of the liver, for example, should avoid uncooked shellfish. They should also avoid high-sodium diets. Patients who have been diagnosed with Wilson's disease should avoid foods that naturally contain copper, including nuts, chocolate, shellfish, and mushrooms.

Diet can affect the urinary system, too. People with kidney disease must be especially careful about the foods that they eat. Even some normally healthy food choices can cause kidney problems to worsen. Doctors recommend that people with kidney conditions limit their daily intake of foods that are high in protein, potassium, cholesterol, and sodium. Unhealthy kidneys can no longer process the protein and potassium properly. Cholesterol and sodium can lead to high blood pressure, a leading cause of kidney failure.

Food and Flatulence

Flatulence is air or gas in the intestines that is expelled from the body through the rectum. Some foods are more likely to give people gas than others. People who are prone to flatulence should avoid the following foods:

- Onions
- Celery
- Carrots
- Beans
- Brussels sprouts
- Raisins
- Bananas
- Prunes
- Bagels
- Pretzels

THE ABCs OF EATING AND FOOD SAFETY

Vitamins and minerals are important to the digestive, urinary, and lymphatic system. Here's how!

- *Vitamin A* keeps the immune system healthy and may help to reduce the risk of esophageal cancer. Foods rich in vitamin A include broccoli, carrots, dairy products, eggs, and spinach.
- *Vitamin B1,* also known as thiamine, helps the body digest carbohydrates. Foods rich in vitamin B1 include whole grains, peas, beans, eggs, and watermelon.
- *Vitamin B2,* also known as riboflavin, helps the body process proteins, fats, and carbohydrates and may reduce the risk of some types of digestive cancer. Foods with this vitamin include eggs, milk, meats, whole grains, and leafy vegetables.
- *Vitamin B3,* also called niacin, helps the digestive system break down food properly and keeps this system healthy. Foods high in vitamin B3 include cheese, beans, milk, meats, poultry, fish, eggs, and nuts.
- *Vitamin B6* helps the body process proteins, fats, and carbohydrates. It also helps the lymphatic system work properly. Foods high in this vitamin include carrots, nuts, bananas, fish, rice, whole grains, and meats.
- *Vitamin B9,* also called folic acid, may reduce the risk of colon polyps. Foods rich in this vitamin include spinach, beans, and fruits.
- *Vitamin B12* helps the body break down foods more effectively. It is found in red meat, eggs, fish, poultry, and dairy products.
- *Vitamin C* aids the body in properly absorbing iron and calcium. It may also reduce the risk of some types of cancer. This vitamin can be found in such citrus fruits as oranges, lemons, and limes, as well as tomatoes, potatoes, and strawberries.

- *Vitamin D* helps the body properly absorb calcium and phosphorous. Foods rich in vitamin D include eggs, milk, and oily fishes. Vitamin D is also produced by skin cells when they are exposed to sunlight.
- *Vitamin K* helps the liver function properly. It is found in egg yolks, fish oils, liver, and green, leafy vegetables.
- *Calcium* may reduce the risk of colon polyps. Calcium-rich foods include milk, cheese, and broccoli.
- *Inositol* (in-OH-sih-tahl) is a mineral that helps the body break down fats and aids in preventing constipation. Inositol is found in citrus fruits, whole grains, raisins, and cabbage.
- *Choline* (KOH-leen) keeps fat from building up in the liver and helps prevent liver and kidney damage. Choline is found in liver, leafy vegetables, and egg yolks.

Biochemist Dr. Casimir Funk, who coined the term *vitamin*, is shown at work in his research laboratory in New York City in 1953.

While there are many nutrients that can maintain and improve health, there are also some nasty organisms that can get into food and cause serious illness. Food-borne diseases affect millions of people each year, causing serious damage to the digestive system, as well as other body systems. Remember to take precautions when handling food that might spread germs. Here are some tips:

- Thoroughly cook all foods—especially red meat and eggs—to kill any bacteria that may be present.
- Carefully wash all cutting boards, knives, and other cooking utensils after each use.
- Always wash your hands before and after handling and preparing food.
- Do not defrost food on the kitchen counter. Instead, use the refrigerator or the microwave.
- Do not marinate food at room temperature.
- Keep hot foods hot and cold foods cold. Use a meat thermometer to make sure that meat and other foods are properly cooked. Always keep the refrigerator at 40°F (4°C) or lower.
- Refrigerate leftovers immediately. Remember the 2-40-140 rule: Don't eat meats, dressings, or sauces that have sat out for more than two hours at temperatures between 40°F and 140°F (4°C and 60°C).
- Wash raw vegetables thoroughly before eating.
- Do not drink or eat foods made from milk that is not pasteurized.
- Avoid cross-contamination, which is the spreading of bacteria from one type of food to another. Keep raw meats and ready-to-eat foods separate from other types of foods.
- Clean the refrigerator regularly. Throw away any foods that are spoiled or old.

Irradiation

Irradiation is the process of zapping foods with X-rays, gamma rays, or electron beams to destroy bacteria. In the United States, this process is used on wheat, potatoes, spices, seasonings, pork, poultry, red meats, fresh fruits, and dehydrated products. According to the Food and Drug Administration (FDA), a federal agency that monitors food safety in the United States, irradiation is perfectly safe. The FDA points out that food irradiation is approved by the World Health Organization (WHO) and is practiced in more than forty nations. However, the FDA warns that irradiation may result in a small loss of the irradiated food's nutritional benefits and that even irradiated food may still carry harmful bacteria and other organisms. The FDA recommends using safe food-handling techniques to avoid disease and illness.

HERE'S TO YOUR HEALTH

Water is an essential part of any healthy diet. The digestive system and the urinary system function more efficiently when a person drinks plenty of fluids each day. Water aids the digestive system by helping it more efficiently break down and move food through the body, preventing constipation. It keeps the urinary system healthy by cleansing the kidneys and flushing bacteria from the urinary tract.

To keep the body as fit as possible, doctors recommend drinking eight to ten glasses of water a day. What happens if you don't have enough water? The body becomes dehydrated, causing headaches and weakness. Severe dehydration can lead to death.

Recent research has even shown that drinking plenty of fluids each day can cut a person's risks of developing bladder cancer. According to the ACS, extra fluids mean more frequent urination. This dilutes cancer-causing substances and empties them out of the bladder People who drink less than 6 cups (1.4 liters) of fluids each day are twice as likely to develop bladder cancer as are those who drink at least 11 cups (2.6 liters) of fluids per day.

Fast Fact

To tell if the food you're buying has been irradiated, look for the words "treated with radiation" or "treated by irradiation" on the label. The package will also carry the international symbol for radiation.

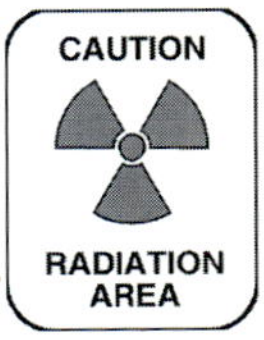

HEALTHY CHOICES

There are a number of choices that people can make to ensure that their bodies work right. One such choice is exercise. Exercise is not just good for the musculoskeletal system. It can keep the digestive system functioning properly by preventing constipation. Daily moderate exercise can even reduce the risks of getting colon cancer and other diseases. It also helps people maintain a weight that is healthy and appropriate for their age, height, and sex.

People with certain health conditions can benefit from exercise. Physical activity increases bone strength, which is important for those undergoing dialysis. It also causes the bones to release more calcium, which reduces the risk of developing kidney stones. People with diabetes should exercise every day to help prevent loss of kidney function.

Because most of the substances that we take into our bodies are processed through the digestive system, harmful products like alcohol, drugs, and cigarettes can cause serious health problems. One of the easiest ways to prolong life and keep the body fit is to avoid these disease-causing substances.

Regular exercise, such as that provided by physical education classes in school, is important in maintaining good health.

Cigarette smoking and the consumption of alcohol is a very unhealthy combination.

Although light alcohol consumption for older people can have positive health benefits, excessive alcohol use or alcohol abuse can lead to serious health problems. Alcohol is the most common cause of liver damage, including hepatitis, cirrhosis, fatty liver, and liver failure. It has also been shown to increase the risk of cancer of the esophagus, pharynx, larynx (LAAR-inks), stomach, colon, rectum, and liver.

Cigarette smoking is extremely toxic to the body. Each year, thousands of Americans die prematurely because they chose to smoke cigarettes. Tobacco smoke has been shown to contribute to polyps, as well as pancreatic and other types of cancer. It has also been shown to irritate and worsen existing conditions, including GERD and gastritis. If you smoke, quit now. The harmful effects of the chemicals and toxins in cigarette smoke will begin to reverse themselves. The longer you stay smoke-free, the healthier you will be.

Illegal drugs such as cocaine, ecstasy, and heroin can cause serious problems to the digestive system, the urinary

Fast Fact

When used together, alcohol and cigarettes can be a deadly combination. The risk of digestive-system cancers and other health problems increases greatly for people who smoke and drink.

system, and the rest of the body. Cocaine can affect the digestive system by causing nausea and abdominal pain. Its effects on other parts of the body, including the circulatory and nervous systems, are even more severe. The so-called club drug methylenedioxymethamphetamine (meth-ih-leen-dye-oks-ee-meth-am-FET-uh-meen), also known as MDMA or ecstasy, can lead to dehydration, liver failure, kidney failure, and death. Using dirty needles to inject such drugs as heroin can lead to hepatitis infection and other serious health conditions, including HIV infection.

Health and Hygiene

Good hygiene plays an important role in preventing disease and illness. Something as simple as washing the hands before and after eating or preparing food can dramatically reduce the risk of ingesting or spreading germs and bad bacteria. According to the FDA's Center for Food Safety and Applied Nutrition (CFSAN), a study examined schoolkids' hygiene habits over a period of thirty-seven days. The study showed that school-age kids who washed their hands four times a day had 24 percent fewer sick days than those who didn't and 51 percent fewer sick days due to upset stomachs.

GLOSSARY

adenoids—clumps of lymphatic tissue, located behind the nose, that trap bacteria and viruses entering the body through the nasal passages

appendix—a portion of the large intestine that has no known function

autoimmune disorder—a condition caused when the body's immune system attacks other parts of the body that it mistakenly identifies as foreign or harmful

bacteria—microscopic, single-celled organisms that are found everywhere; many bacteria cause disease, but not all are harmful

benign—not cancerous

biopsy—a surgical procedure to remove tissue, cells, or fluids from the body for examination

birth defect—any physical problem with which an infant is born

bladder—a balloon-shaped, muscular organ that stores urine until it is ready to be passed from the body

bone marrow—a type of tissue found in the center of many long bones where infection-fighting cells are produced

cancer—a disease in which a group of abnormal cells in the body divides without control

chemotherapy—the use of anticancer drugs to cure or stop the spread of cancers

cirrhosis—a medical condition in which the liver becomes scarred and damaged

computerized tomography (CT)—a test that uses a thin beam of electromagnetic radiation to create a three-dimensional picture of what is inside the body

constipation—a condition that occurs when the muscles of the large intestine move too slowly, causing stool to absorb too much water and become hard

dialysis—a medical procedure that removes toxic substances and wastes from the body

diarrhea—loose, watery stools that occur more than three times in one day, often caused by food-borne or water-borne bacteria, viruses, and parasites

digestive system—the body system responsible for turning food into substances that can be used by the body for energy and nourishment

eating disorder—a condition in which a person has a distorted image of his or her body and uses food in ways that harms the body

end-stage renal disease (ESRD)—a serious, irreversible medical condition in which kidney function drops below 20 percent

environmental health—the body's reaction to conditions that people have little or no control over, including sunlight, bacteria, chemicals, and pollution; it also includes medical problems that result from personal lifestyle choices, such as smoking

esophagus—a muscular tube that connects the throat to the stomach

food-borne disease—an illness transmitted to humans through food or water

fungi—parasitic organisms, such as mushrooms and molds, that feed on both living and decaying matter

gallbladder—a sac in the digestive system that stores bile

gallstones—rocklike pieces that form in the gallbladder when bile hardens, causing health problems

gastroesophageal reflux disease (GERD)—a condition in which the valve between the stomach and esophagus doesn't close properly, resulting in stomach acids escaping into the esophagus; sometimes known as *heartburn*

genetic disorder—an abnormality in a person's genes or chromosomes

hepatitis—an inflammation of the liver that may be caused by viruses, autoimmune disorders, or other medical conditions

hernia—a rupture or tear in the lining of a muscle, tissue, or membrane that holds an organ in place

human immunodeficiency virus (HIV)—a contagious virus that attacks the body's immune system; HIV can lead to AIDS

kidneys—fist-sized, bean-shaped organs that filter the blood to remove impurities and excess water

large intestine—also known as the colon or bowel, this part of the digestive system is the last section to process food before it is expelled from the body

liver—a digestive organ that makes bile and helps the body use food

lymph—clear fluid made up of water, proteins, salts, and other substances that carries excess body liquids away from the body's cells and tissues

lymphatic system—the body system responsible for filtering excess water, proteins, and fat from the body's cells and tissues and returning them to the blood vessels

lymphedema—a condition in which the lymphatic vessels swell and become blocked, causing lymph to build up in the fatty tissue just below the skin

lymph nodes—small, bean-shaped organs found throughout the body that filter harmful organisms out of the blood and produce infection-fighting cells

lymphoma—a cancer that begins in the lymphatic system

lymph vessels—tiny vessels, similar in structure to veins and capillaries, that connect lymph nodes to each other

magnetic resonance imaging (MRI)—a test that uses magnets and radio waves to create an image of the inside of the body

malignant—cancerous or harmful

pancreas—a digestive organ that produces enzymes used in breaking down food

radiation—energy in the form of rays or particles

renal function—the workings of the kidneys

small intestine—a part of the digestive system that absorbs nutrients from food into the bloodstream

spleen—an organ that filters damaged cells from the bloodstream, destroys them, and sends them to be cleared out of the body

stomach—a pouchlike organ that receives, holds, and partly digests food

thymus gland—a small organ near the base of the neck where some types of infection-fighting cells mature

tonsils—two clumps of lymphatic tissue, located in the throat, that trap bacteria and viruses entering the body through the mouth

transfusion—a medical procedure in which blood or another liquid is transferred into an artery or a vein

tumor—an abnormal mass of tissue that grows in the body; tumors may be benign or malignant

urea—liquid waste that circulates in the bloodstream and is removed from the body by the urinary system

ureters—tubes that carry urine from the kidneys to the bladder

urethra—a tube through which urine passes from the bladder out of the body

urinary stones—stones that form in any part of the urinary system, usually as a result of excess salts and minerals in the body

urinary system—the body system responsible for filtering excess water and liquid waste from the bloodstream and expelling it from the body

villi—small, fingerlike organs that line the intestine and absorb nutrients from food to transmit to the bloodstream

virus—a simple germ organism that is always harmful to humans; viruses are found everywhere

X-ray—a test that uses electromagnetic radiation to take a two-dimensional picture of what is inside the body

BIBLIOGRAPHY

BOOKS

Avraham, Regina. *The Digestive System.* Philadelphia: Chelsea House, 2000.

Clayman, Charles, ed. *The American Medical Association Encyclopedia of Medicine.* New York: Random House, 1989.

———. *The Human Body: An Illustrated Guide to Its Structure, Function, and Disorders.* New York: DK Publishing, 1995.

Edelson, Edward. *The Immune System.* Philadelphia: Chelsea House, 1999.

Kittredge, Mary. *The Human Body: An Overview.* Philadelphia: Chelsea House, 2001.

Silverstein, Alvin, Virginia Silverstein, and Robert Silverstein. *The Excretory System.* New York: Twenty-First Century Books, 1997.

Toriello, James. *The Stomach: Learning How We Digest.* New York: Rosen Publishing Group, 2002.

Walker, Richard, ed. *Encyclopedia of the Human Body.* New York: DK Publishing, 2002.

White, Katherine. *Everything You Need to Know about AIDS and HIV.* New York: Rosen Publishing Group, 2001.

WEB SITES

CDC National Center for Environmental Health (NCEH)
www.cdc.gov/nceh

The Children's Environmental Health Network
www.cehn.org/cehn

KidsHealth
kidshealth.org

The Leukemia and Lymphoma Society
www.leukemia-lymphoma.org

National Eating Disorders Association
www.nationaleatingdisorders.org

The National Lymphedema Network (NLN)
www.lymphnet.org

NIH National Institute of Diabetes and Digestive and Kidney Disorders (NIDDK)
www.niddk.nih.gov

NIH National Institute on Drug Abuse (NIDA)
www.nida.nih.gov

United States National Library of Medicine
medlineplus.nlm.nih.gov

INDEX

Note: Page numbers in *italics* refer to maps, illustrations, or photographs.